PRAMOD BELKHODE
PRATIBHA AGRAWAL
ADITYA CHAVHAN, SARVESH SAHA

MATERIAL SCIENCE ENGINEERING

AF792201

PRAMOD BELKHODE
PRATIBHA AGRAWAL
ADITYA CHAVHAN, SARVESH SAHA

MATERIAL SCIENCE ENGINEERING

STRUCTURE AND SUBSTANCE : THE SCIENCE OF MATERIAL ENGINEERING

LAP LAMBERT Academic Publishing

Imprint
Any brand names and product names mentioned in this book are subject to trademark, brand or patent protection and are trademarks or registered trademarks of their respective holders. The use of brand names, product names, common names, trade names, product descriptions etc. even without a particular marking in this work is in no way to be construed to mean that such names may be regarded as unrestricted in respect of trademark and brand protection legislation and could thus be used by anyone.

Cover image: www.ingimage.com

Publisher:
LAP LAMBERT Academic Publishing
is a trademark of
Dodo Books Indian Ocean Ltd. and OmniScriptum S.R.L publishing group

120 High Road, East Finchley, London, N2 9ED, United Kingdom
Str. Armeneasca 28/1, office 1, Chisinau MD-2012, Republic of Moldova, Europe
Managing Directors: Ieva Konstantinova, Victoria Ursu
info@omniscriptum.com

Printed at: see last page
ISBN: 978-620-8-06381-8

Copyright © PRAMOD BELKHODE, PRATIBHA AGRAWAL, ADITYA CHAVHAN, SARVESH SAHA
Copyright © 2025 Dodo Books Indian Ocean Ltd. and OmniScriptum S.R.L publishing group

OVERVIEW

The Book of Materials offers a comprehensive and detailed exploration of the fundamental concepts of materials science, essential for students and professionals alike. It begins with an Introduction to Materials, where readers are introduced to the various classes of materials, including metals, ceramics, polymers, and composites. This section highlights the unique properties of each material type, explaining how these properties determine their applications in industries ranging from aerospace to biomedical engineering. The book then delves into Crystal Structure, an in-depth examination of how atoms are arranged in materials. Concepts such as unit cells, crystal lattices, and crystallography are discussed to show how atomic arrangements influence key material properties such as strength, hardness, and conductivity. The book further explores the Imperfections in Solids, focusing on defects within the crystal structure, including vacancies, dislocations, and grain boundaries. These imperfections play a critical role in determining the mechanical, electrical, and thermal properties of materials, impacting how they perform in real-world applications. The next chapter, Derivation of Material Properties, connects these structural concepts to the physical and mechanical behaviors of materials. This section explains how properties like tensile strength, elasticity, and thermal conductivity are derived from atomic and molecular structures, providing a deeper understanding of how materials behave under stress, heat, or other environmental conditions. The final chapter, Nanocomposite Materials, explores one of the most cutting-edge areas in materials science. This chapter details the integration of nanoparticles into traditional materials to create nanocomposites, which offer enhanced properties such as increased strength, lighter weight, and improved electrical conductivity. Applications in fields such as electronics, energy storage, and medicine are highlighted, showing how nanocomposites are revolutionizing material design. Overall, the Book of Materials presents a thorough and nuanced understanding of how the structure of materials directly influences their properties and applications, making it an indispensable resource for advancing knowledge in this ever-evolving field.

TABLE OF CONTENT

CHAPTER 1

CHAPTER 2

CHAPTER 3

CHAPTER 4

CHAPTER 5

CHAPTER 6

CHAPTER 7

CHAPTER - I

Introduction to Materials

1.1 Introduction to Materials

Bonding between atoms: metallic bonding, ionic, covalent and vander waal interaction.

Thermal expansion, elastic modulus and melting point of materials, role of materials selection in design, structure property processing performance relationship.

Introduction of Materials-

Since the 1960, the term material science and engineering has come in branches of engineering related with materials which include metallurgy, ceramic engineering, polymer chemistry, physics, and physical chemistry.

The word science describes atomic bonds, crystalline structure, crystal defects and non-crystalline structure, mechanical behavior (elastic and plastic behavior), Thermal behavior, thermal expansion, thermal conductivity, failure analysis impact energy, non- destructive test, phase diagram, phase rule, eutectic, phase solid solution, kinetics hardening, annealing.

The word material describes metals, ceramics, glasses, polymers, composites, electronic and magnetic materials, electrical behavior, optical behavior, semiconductor and magnetic materials.

The word engineering describes environment, degradation and material selection. We live in the world of materials which relates to our social relationship and economical quality of life.

Material with which earlier ancestor where probably using for their tools and weapons are-

Stone age- Almost 2.5 million years ago when human ancestor used stone to form weapons for hunting.

Copper age- Roughly 4000 BC, relatively pure copper where used, existed in Europe.

Bronze Age- Period 2000 BC to 1000BC- foundation of metallurgy in which alloys of copper and tin were discovered to produce superior tools and weapons.

Iron age- 1000BC to 1BC by 500BC, iron alloys had largely replaced bronze for tools and weapons making in Europe.

Pottery age- Domestic vessels made from clay glass artifacts that is solid piece have been traced back to 4000 BC.

Modern Culture 20th century deferred to as plastic material which are lightweight economical and polymeric materials.

Since 1960, pressure for weight and cost savings have led to an increasing demand for new non- metallic materials specially automobile materials.

1.2 Semiconductor-

Here we are having a circuit with a battery a light bulb and a gap.

Fig.1

If I connect metallic nut which will be conductor then the light will glow; if we connect glass bottle then it will be insulator and hence no light will glow; and if silicon layer or wafer which is a thin slice of silica is attached, then the light stays off because it is a insulator. When the Heat is passed, again the light glows. Silicon is insulator at room temperature but conducts electricity when it becomes very hot. It is semiconductor, whose conductivity changes based on the other factors. This special ability makes semiconductors the perfect brain for electronic devices. Circuits of small semiconductor switches, called transistor are heart of computer chips which help to do maths and run programs.

Materials are of three types

1.conductor

2.insulator and

3.semiconductors

Electrons are distributed randomly in the conductor. As the temperature increases randomness increases and that's why the current decreases, but in case of semiconductors as the temperature increases current increases. Semiconductors have enabled electronics to become smaller, faster, and more reliable. Semiconductors- which allow them to either conduct or insulate in a single atom, electrons can occupy specific energy levels. When multi atoms bond, the electrons are shared energy level shift around. Solids, trillions and trillions of atoms interact with each other.

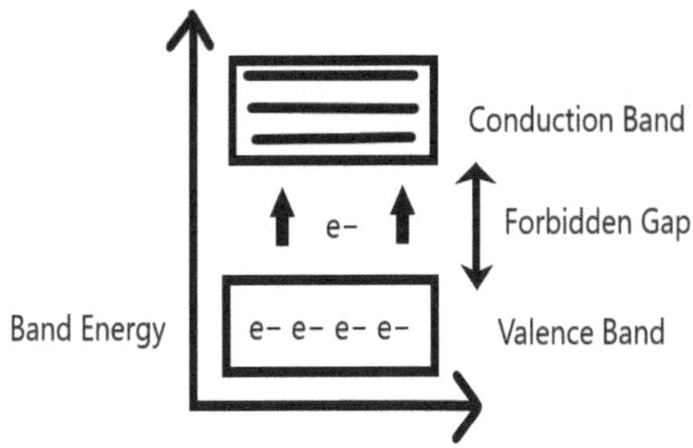

Fig. 1.2

Semiconductors fall somewhere in the middle, have medium size band gap. If glass is heated, it breaks.

Transistor make up the logic elements, the memory component and communication modules that computers talk to each other. Almost in all devices, spaceship, it serves in making the instruments. If there is no semiconductors then there will be no information age.

Atoms - Atom is the smallest unit of any substance. Atoms are so small that it cannot be seen by naked eyes. Sugar contains atoms of carbon, hydrogen and oxygen; salt contains atoms of sodium and chlorine; water contains atoms of oxygen and hydrogen. Same atoms combine to form elements, when different atoms combine, it forms compounds. For example, sodium and chlorine forms salt. Water contains oxygen and hydrogen. When millions of hydrogen and oxygen combined together chemically, it can be seen with electronic microscope.

Atom contains proton neutron and Electrons. Atoms are different due to the different number of protons, electrons, and neutrons. E.g. oxygen atom contains eight proton, eight electrons and eight neutrons. Carbon atom contains six protons full, 6 electrons and six neutrons.

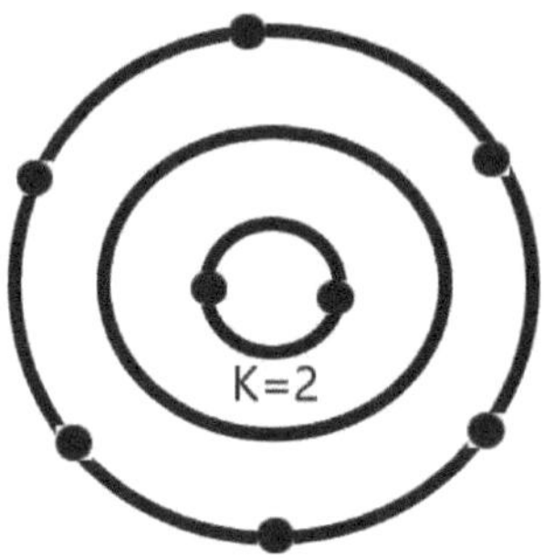

Fig1.3

1.3 Atomic bonding-

Scanning microscope allows the imaging of individual atoms bonded to a material surface.

Primary Bonding	Secondary Bonding
Bond Energy 100-1000 kJ/mol	Bond Energy 1-50 kJ/mol
It involves the transfer and sharing of Electrons and produces a relatively strong joining of adjacent atoms.	It involves a relatively weak attraction between atoms in which no electron transfer or sharing occurs.
Ionic, covalent and metallic bonds.	Vander waal bonds.
Bond Length 1-2A0/ 0.1-0.2 nm	Bond Length 2-5A0/ 0.2-0.5 nm

Fundamental engineering materials like metals commerce ceramics, glasses, polymers, semiconductors all are associated with the various types of atomic bonding for example composites are combination of fundamental types metals shows metallic bonding; ceramics shows covalent and ionic bonding; glass shows ionic bonds present between positive and negative ions; polymers show covalent bonds and secondary bonding, and composite shows various types of adhesive bonding.

1.4 Atomic Structure

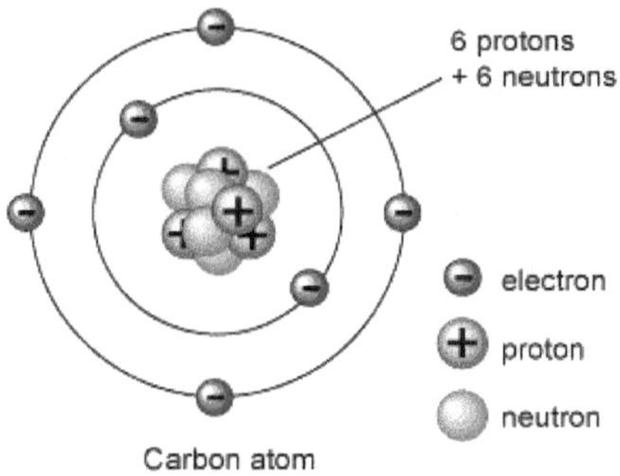

Fig 1.4

(Schematic model of carbon atoms-)

Chemical identification is done relative to the nucleus, atomic bonding involves electrons and electron orbitals.

Bonding of adjacent atoms- strong primary bonds are formed when outer orbit related bonds are transferred or shared between atoms. Vicar secondary bonds results from a more delicate attraction between positive and negative charges with no actual transfer or sharing of Electrons.

The various possibilities of bonding in a systematic way

Ionic bond,

Covalent bond,

Metallic bond,

Vander Waal's forces.

Ionic Bond- Ionic bond is the result of electron transfer from one atom to another. It has been observed by Charles Augustin Coulombic (1736), a French Scientist, who demonstrated that the force of attractions contribute in electricity, magnetism, friction.

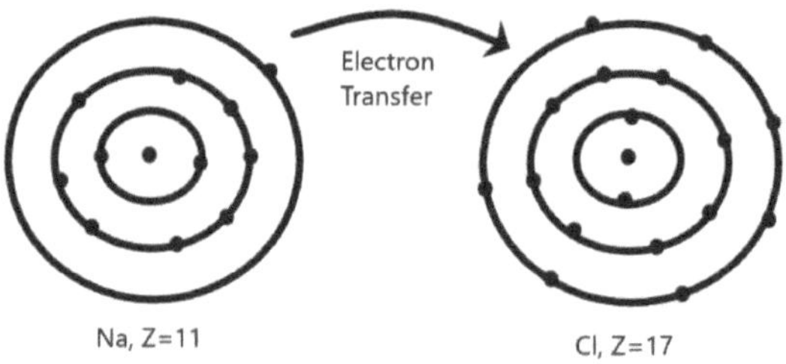

Fig 1.5

(Ionic bonding between sodium and chlorine atoms.)

Electron transfer from sodium to chlorine, creates a cation Na+ and an anion Cl-. The ionic bond is due to the Coulombic attraction between the ions of opposite charge. The transfer of an electron from sodium is favored because it produces a more stable electronic configuration that is resulting in Na+; similarly, the chlorine readily accepts the electron full, producing a stable Cl- ion having full outer orbital shell. The charged species Na+ and Cl- are termed as cation and anion respectively. Ionic bond is non- directional. A positively charged Na+ will attract any adjacent Cl- equally in all directions. Figure shows how any positive and negative ions are stacked together in solid sodium chloride which is known as rock salt.

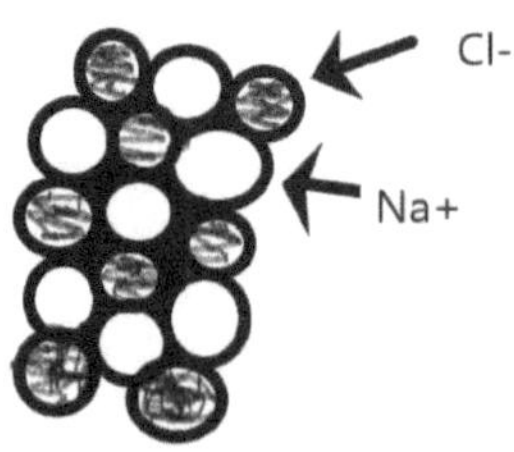

Fig 1.6

(Ionic bond is a result of Coulombic attraction between oppositely charged species.)

Fc is coulombic force of attraction between two oppositely charged ions, a is the separation distance between the centres of ions, $K=Ko(Z_1q)(Z_2q)$; Z is the valence of the charged ions.

(eg: +1 for Na+ and -1 for Cl-, q= charge of single electron=0.16x 10 -18C, Ko is proportionality constant= 9 x 10 9 Vm/C).

Figure shows that the Columbian force of attraction increases as the separation distance between adjacent ion centers decreases; this in turn implies the bond length a. Bond length are not zero because the attempt to move two opposite charge ions closer together- to increase the Coulombic attraction is contracted by opposing repulsive force Fr which is due to the overlapping of the similarly charged electric fields from each ions as well as attempt to bring two positively charged nuclei closer together. Repulsive force Fr = Ke (q^2 / d^2)

where Lambda and rho are constant for ion pairs.

Net bonding force F= Fc+ FR.

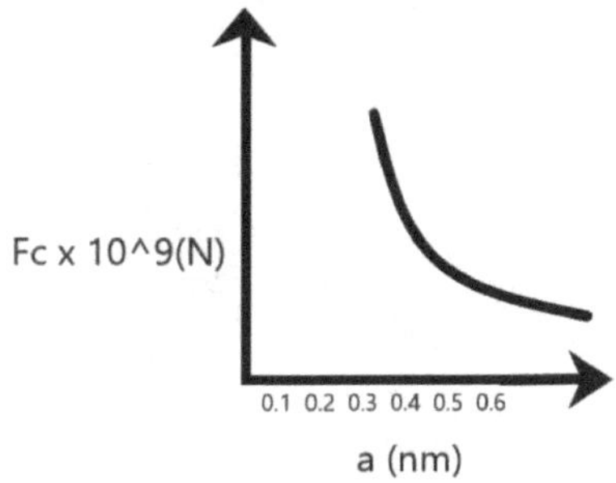

Fig 1.7

The equilibrium bond length a0 occurs at the point where the force of attraction and repulsion are balanced that is 0= Fc+ FR.

Externally applied Compressive force and tensile force to push and pull ions indicate mechanical behavior of solids.

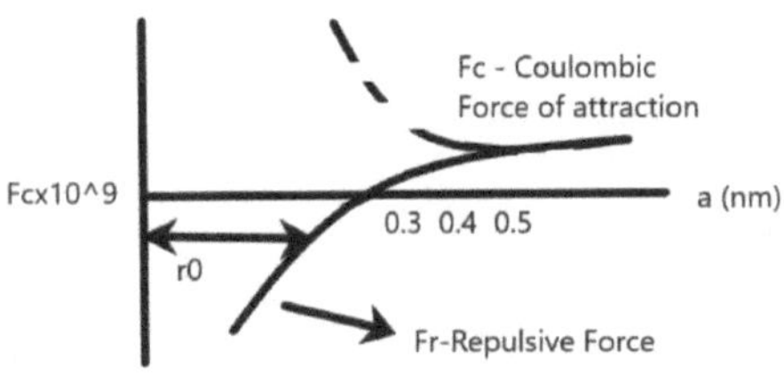

Fig1.8

1.5 Coordination Number-

Figure shows a structure for NaCl in which six Na+ surround each Cl- and vice versa. The coordination number C N is the number of adjacent ions or atoms surrounding a reference ion or atom. For each ion shown in the figure, the coordination number is equal to 6 that is each ion has six nearest neighbors. For ionic compounds, the coordination number of the smaller ion can be calculated systematically by considering the greatest number of larger ions of opposite charge that can be in contact with, or coordinate with the smaller one. This coordination number depends directly on the relative sizes of the oppositely charged ions this

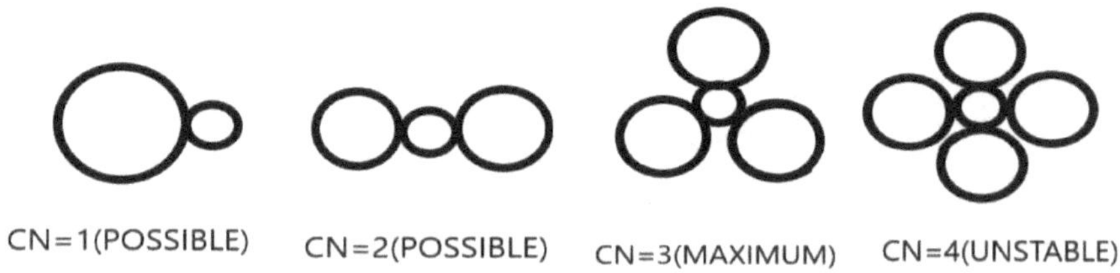

Fig 1.9

relative size is characterized by the radius ratio r/ R, where r is the radius of the smaller ion and R is the radius of the larger ion.

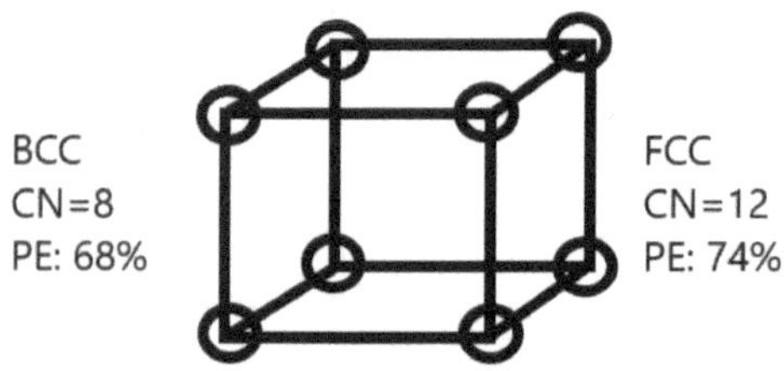

Fig 1.10

The coordination number is the number of bonds the central metal atom is bonded to. Now in the complex, coordination number can help determine the name of coordination complex.

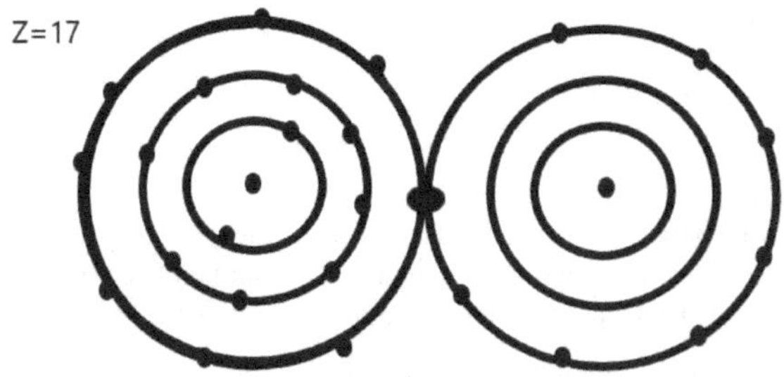

A PLANETARY MODEL

Fig 1.11

The covalent bond – This is highly directional the name covalent derives from the cooperative sheet of valence electrons between two adjacent atoms. Valence electrons are those outer orbital electrons that involves in bonding. Figure shows that covalent bond in a molecule of a chlorine gas with

-a planetary model

-the actual electron density

-an electron dot schematic diagram and

-a bond line schematic diagram.

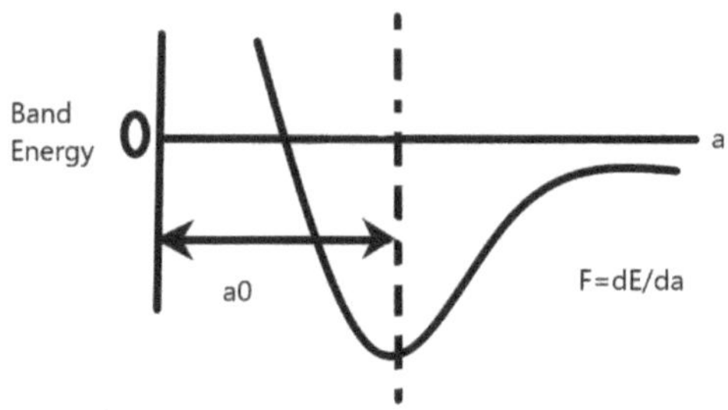

Fig 1.12

A single line indicates a bond between two atoms that involves one electron pair; a double bond is indicated by double lines involving two electron pairs; a triple bond represents sharing of three bonding pairs which ensure the atomic stability in their outer shell.

The idea that two electrons can be shared between two atoms and serve as the bond between them was first introduced in 1916 by American chemist G N Lewis. Covalent bonds are directional which means that the atoms bonded prefer specific orientation relative to one another. This in turn gives molecules a definite shape. Electronegativity is the tendency to attract electrons or it is the charge with negative electricity.

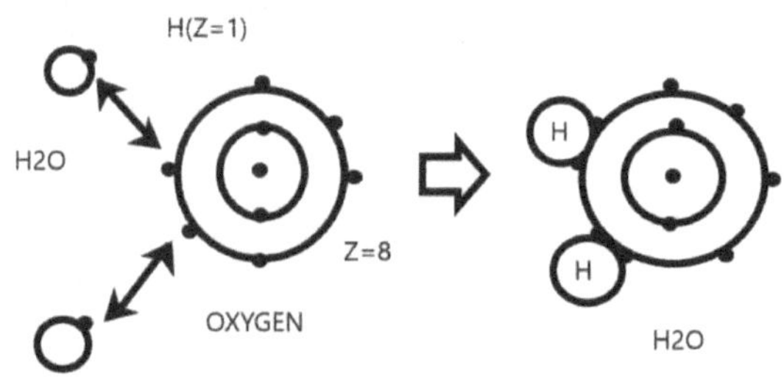

Fig 1.13

Metallic bonding- Metals are used in many objects in everyday life. For example, the pot used for the cup of tea is made up of metal. The kettle is made up of plastic but the coil which is the heating element inside, is made up of metal. Metals are good conductors of heat. Metals are used to make these everyday objects. Also, they are conductor of electricity. The wire connecting your kettle to the electrical socket is made of many copper wires insulated with a layer of rubber. Pot, Kettle, and Copper Wire all have different shapes. It's malleable. It means that they can be molded into different shapes. Also, they are ductile. This means that they can be stretched into wires. To understand the properties of metal, we have to understand the metallic bonding.

The ionic bond involves electron transfer and is non- directional. The covalent bond involves electron sharing and its directional. Metallic bond involves electron sharing and is non-directional.

In this case the valence electrons are delocalized electrons which have an equal probability of being associated with any of large number of adjacent atoms leading to electron cloud or electron sea or electron gas. Mobile gas is the basis of high electrical conductivity in metals. Metallic elements tend to have lower value of electronegativity than non- metals.

It describes the electrostatic attraction between the metal ions, arranged in a lattice structure, and the free-floating electrons around them. Since these electrons are free to move around, the term sea of electrons is also used. Metallic structures have only metal ions. These metal ions are arranged side by side in a regular repeating pattern. The free-floating electrons act like glue and hold the structure in place. This is a very strong attraction and metals have high melting and boiling points. A lot of heat energy is needed to overcome this attraction. This shows that metals are very good conductors of heat. Free-flowing electrons are responsible for metals to conduct electricity. The strong electrostatic attraction will remain and therefore structure stays intact.

Metals are malleable and ductile because no matter what the shape of the metal takes place, the free-floating electrons will conform to that shape. In metallic bonds, the smallest unit is an atom like Fe with no transfer of electrons and no sharing of electrons. Iron atoms are joined chemically by electron sea theory.

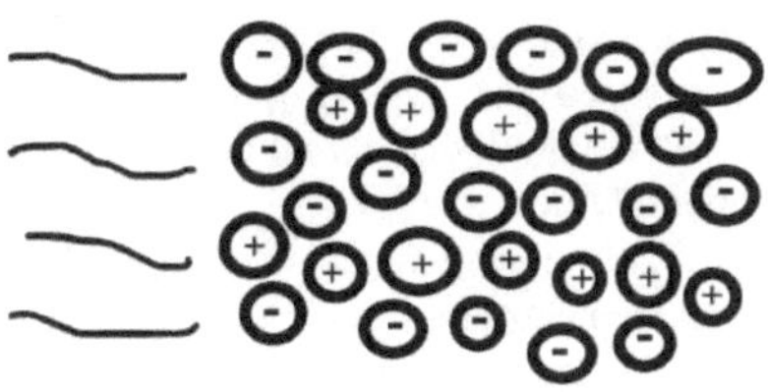

Fig 1.14

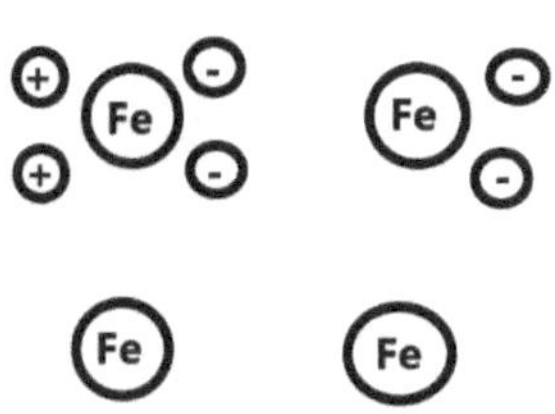

Fig 1.15

Iron Sheet- The iron atom lost two electrons and becomes double positively charged. Another fee atom loss two electron and become double positively charged. Same with other so this process of losing valence electron is continued until all the atoms of the sheet will lose the valence electrons and become positively charged. So, the structure of iron sheet as shown in figure after losing electrons.

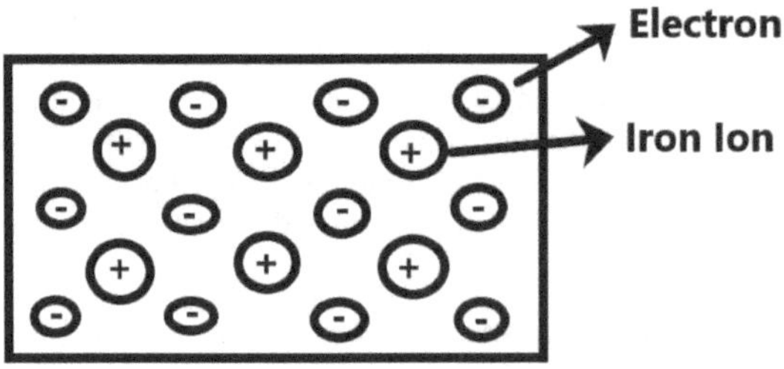

Fig 1.16

It shows that positive charge is iron ion which loss electrons and will form a sea of Electrons Cloud around the positive ions. We know that positive and negative charges attract each other due to electrostatic force. There is a prosthetic force between all positive ions and electronegative sea which will acts as a metallic bond and hold together all the atoms of iron. Note down that is lost electrons do not belong to any individual atoms rather they form a cloud or sea of Electrons. Thus, electrons are called Delocalized because they move freely within the ion sheets so they are also good conductor of heat and electricity. Metallic bond exist in all metals like zinc, copper, gold etc. It is formed due to electrostatic force of attraction between sea of Electrons and positive ions.

Vander Waal's bond- Binding energy is very less amount that is 0.1 electron volt. Melting and boiling point is very low. Conductivity, poor conductor of heat and electricity has outer shell is fulfilled.

Temperature dipole- Consider neutral atom. Electrons are equally distributed electrons are constantly revolving around the nucleus.

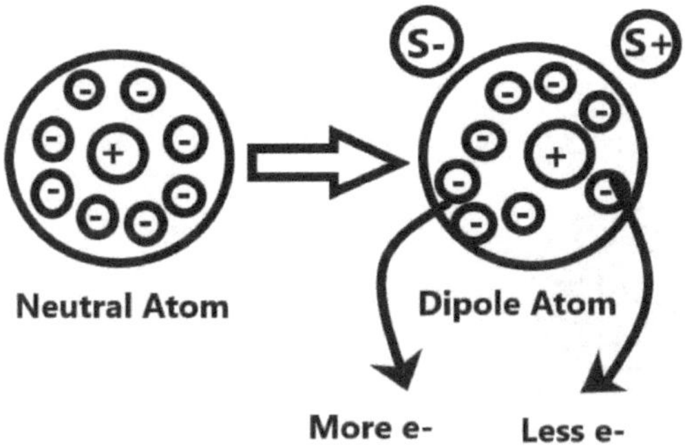

Fig 1.17

There will be time when one side of atom will have more electrons than the other and structure of atoms will be distorted as shown in figure. Here at one side of atom, there is more electrons and at the other side of atom there is less electron or electron deficient. As a result of this, this side becomes partially negative because it contains more electrons. Another side becomes partially positive because it contains less electron. Now this atom is a dipole because it has two positive poles and two negative poles. There will be time when these electrons could be again equally distributed and the positive and negative poles will be vanished and this dipole will again become neutral atom. This atom instantly become a dipole and instantly becomes a neutral atom and therefore this is known as **temporary dipole.**

Temporary dipole is a dipole when the electronic cloud of a neutral atom is distorted for the time being.

Induced dipole- Consider temporary dipole. Temporary dipole has temporary positive pole and temporary negative pole. Now, bring neutral atom near to this temporary dipole. What you observe that, appositive side of temporary dipole will pull the evenly distributed electrons. Neutral atom has a distorted set of electrons now. Atom becomes polarized. This side of electron reach is partially negative and the other side having less electrons is the electron deficient side which is partially positive.

Here two polarized dipoles deposit are temporary dipole and induced dipole. Now positive end of the temporary diaper attracts the negative end of induced dipole. Force between this polar molecule is called as van der Waal forces which are weak and intermolecular forces between particles. These forces are the attractive forces between temporary dipole and induced dipole. It exists in polar molecules like hydrogen fluoride and non polar molecules like CO2. At low temperature molecular solid gas into liquid state or solid state. At room temperature very weak

force means crystal structure in the form of gasses not solid. Because temperature under all force dominate the thermal force.

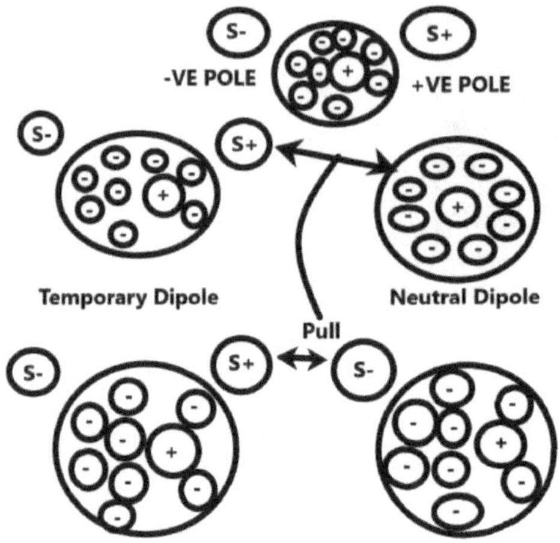
S-
S+
-VE POLE
+VE POLE
S-
S+
Temporary Dipole
Neutral Dipole
Pull
S-
S+
S-

Fig 1.18

CHAPTER 2

Crystal Structure

Miller indices of planes and directions, packing of atoms inside solids, close-packed structure, the structure of ceramics, ionic solids, glass, and polymers, and the density of various materials.

2.1 Introduction –

Crystals are solid materials where atoms and molecules are arranged in a periodic ordered structured manner called a crystal lattice. The study of crystallography is important to understand the structure-property relationship of various classes of crystalline materials. Processing of materials often results in the generation of various crystallographic defects such as dislocation, stacking faults, etc. present in the material which affects the mechanical, optical, and electrical properties of materials and hence, the performance of devices fabricated; based on these materials. Crystallography of a given material is important before examining its various physical properties.

Experimentally, crystallography is studied by performing X-ray and electron diffraction measurements on real materials The X-ray diffraction technique is the basis on which one can examine the structure of materials. The electron diffraction technique is more powerful and used to examine at nano-length scales and to examine the various defects present in the material.

Various properties of metals are governed by their crystal structures. Broadly speaking, the crystal structure is a definite arrangement of atoms and is also referred as atomic structure of metals. All metals and their crystals are crystalline solids.

Crystal structures are studied by their atomic arrangements in particular planes and directions. The density of metal depends upon packing factor of atoms. The plastic deformation property of metals depends upon the number of atoms in a typical plane. Plastic deformation takes place by slip, twinning, or both. Metals are deformed by hot working or cold working.

To analyze the structure of crystals, the Miller system shows how the faces of crystalline structures are placed in relation to each other. This helps to determine the molecular basis and structure of substance and makes cutting or forming the crystals easier and more accurately.

2.2 Crystal structure

Crystal- A crystal is defined as a solid in which definite arrangement of atoms is observed in three directions. Due to fixed position of atoms a crystalline solid is considered as a 3-dimensional network.

Unit cell- It is defined as a smallest arrangement of atoms which is the true representation of the crystal structure and repeats in all possible directions.

Space lattice- It is defined as a three- dimensional network of imaginary lines connecting the atoms.

Lattice parameters- The distance between the center of two atoms is defined as lattice parameters. It is also defined as length of unit cell.

Lattice angles- These are defined as the angles between the edges of a unit cell. Figure shows a typical unit cell. In this unit cell, a, b and c are the lattice parameters while, α, β and γ are the lattice angles. This unit cell is situated at the origin of crystallographic axes x, y and z mutually perpendicular to each other.

The crystal systems- The various crystal systems that are commonly observed are

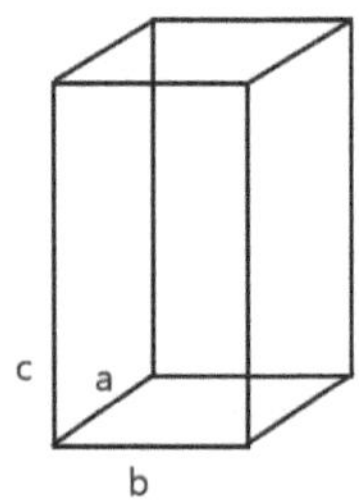

Fig 2.1

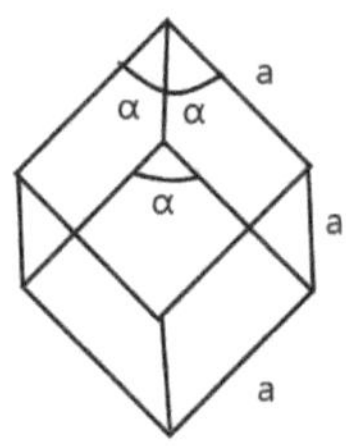

Fig 2.2

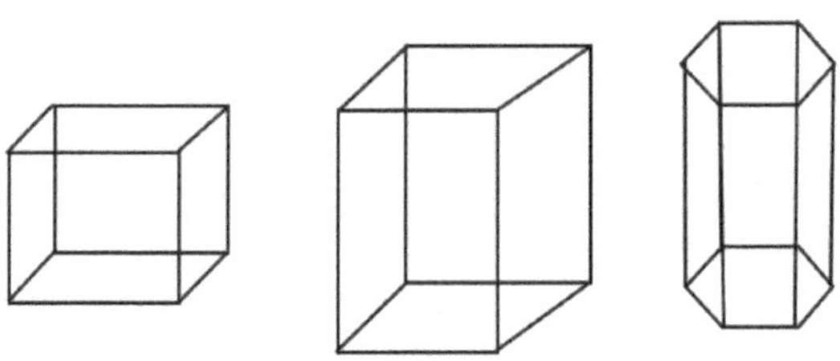

Fig 2.3

Most of the crystal systems are symmetrical in geometry, most unsymmetrical crystal systems are triclinic. The axial distance and axial angle are different. The highest symmetry is observed in cubic systems. The lattice parameters a, b, and c are unit cell edge lengths, and the lattice parameters α, β, and γ are angles between adjacent unit cell access where α is the angle viewed along a axis means the angle between b and c axis.

A lattice is an array of points with identical surroundings in three-dimensional space. These lattices are skeletons upon which crystal structures are built by placing atoms or groups of atoms on or near the lattice points.

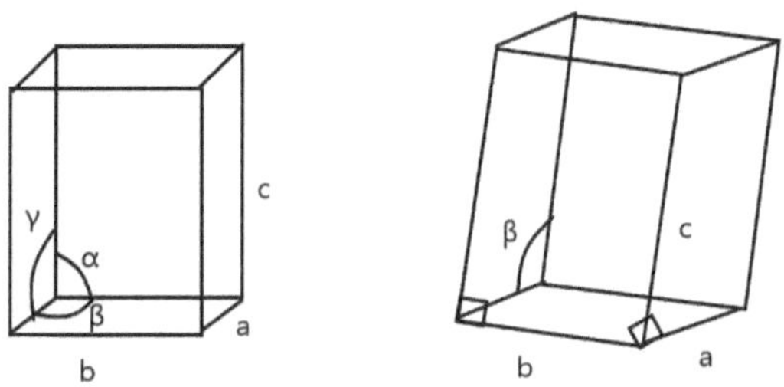

fig 2.4

2.3 Miller Indices of Planes and Directions

Miller indices are a symbolic vector representation for the orientation of an atomic plane in a crystal lattice and are defined as the reciprocal of the fractional intercepts that the plane makes with the crystallographic axis. Miller indices were introduced in 1839 by the British Mineralogist, William Miller, although an almost identical system that is Weiss parameters had already been used by a German mineralogist Christian Weiss since 1817.

Weiss Indices	Miller Indices
Planes are represented by x, y, and Z.	Reciprocal of Weiss Indices
Intercepts of those planes on three axes are Weiss Indices.	
Weiss indices have fractional as well as infinite values.	Expressed as the smallest whole number

Miller indices is used to specify the orientation of plane or direction of the crystal lattice. It is denoted by (hkl). It is important that hkl is presented in which bracket?

( )- round bracket represents Miller indices of plane

[ ]- square bracket represents Miller indices of direction

{ }- curly bracket represents set of symmetrical planes

<   > angle bracket represents set of symmetrical direction is angular.

2.4 Purpose of Miller indices-

For one crystal lattice, all the lattice points from origin is having different directions. Thus, to specify the direction of the lattice point with respect to origin or reference, Miller indices are used.

Oone lattice is consist of different planes, all the planes have different orientation. In such case, to specify the orientation of planes, Miller indices are used.

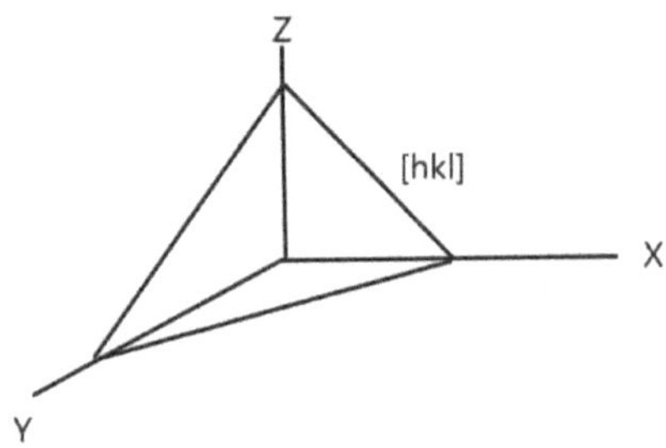

Fig 2.5

To specify the lattice plane or direction of lattice point, how many indices are used? It depends on the dimensions of the lattice.

3D cubic lattice-

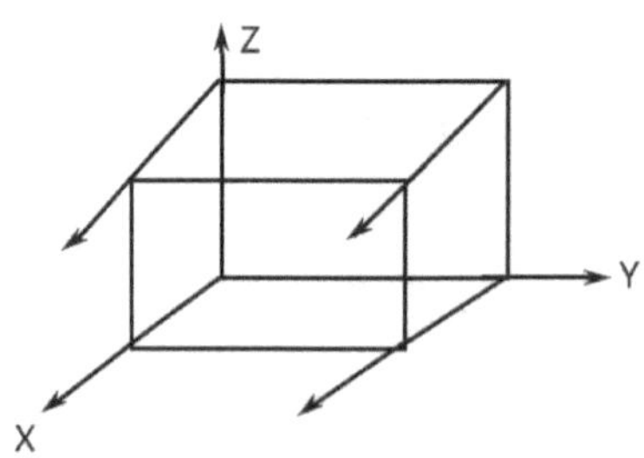

Fig 2.6

Miller indices show different direction, but magnitude is same; so both [100], [010] the Miller indices are in one family member.

In addition to this [100],[010], there are four more members in family.

Family of directions - members in family-

2.5 Miller indices of directions-

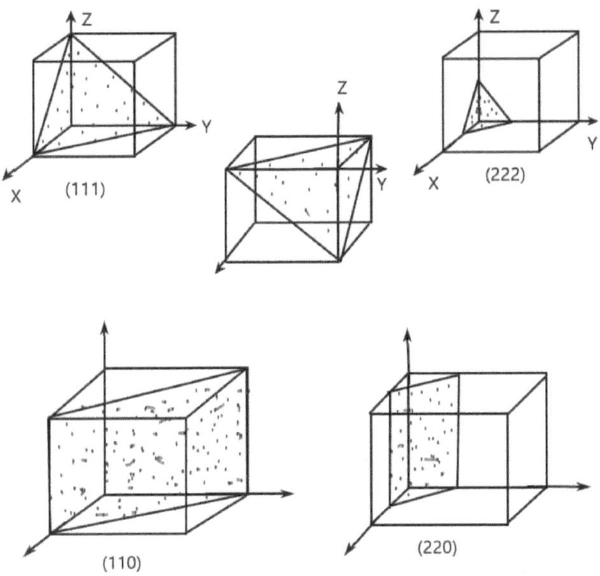

Fig 2.7

Many other directions parallel to PQ

Miller Indices is having great advantages for all parallel directions, Miller indices are same, no need to separate Miller indices.

Steps for Miller indexes for direction-

1. Choose a point on the direct direction as the origin
2. You are free to choose origin anywhere, only origin should be on direction
3. Choose the coordinate system [x,y,z direction]
4. Find the coordinate of another point on the direction in terms of a,b and c 1,0,0
   a,b,c are the three lattice parameters of unit cell a along x- axis;a=1
   Miller indices of dark line is [1a+0b+0c] to represent this direction.

5. Freedom of origin if had parallel lines, all parallel direction have the same Miller indices.
6. Deformation under loading (slip) occur in certain plane of direction.
7. Mode of failure can be predicted.
8. The properties can vary in a crystal with orientation.
9. Reduce the coordinate to smallest integers 1,0,0. Multiply by common factor so that fraction get cancelled.
10. Put in square bracket [100] without commas.
11. For example

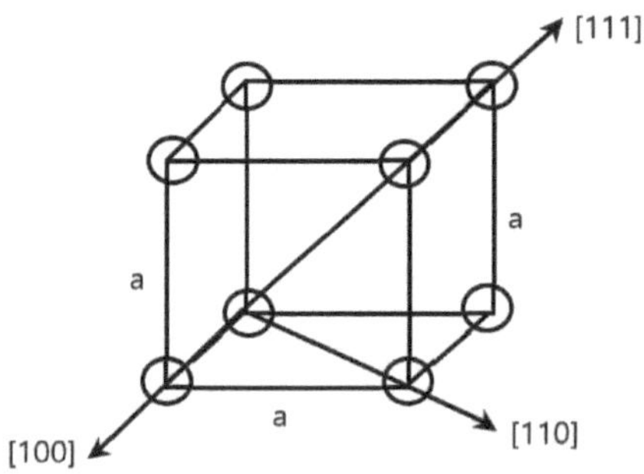

Fig 2.8

Miller indices of planes-

1. Select the origin- select the origin nearest corner to plain and not passing through the given plane.
2. Find the intercept- if plane is parallel to any axis, then intercept is infinity.
3. Form the reciprocal- reciprocal of Infinity is zero.
4. Reduce the fraction- by multiplying with common factors.
5. Enclose the whole number (   ), parenthesis without common.
6. If Miller indices comes out negative; for example -5 then we will write as 5

Step 1-Select origin nearest corner to plane not passing through the given plane

Step 2 - Find intercept

Miller indices used for –

For calculation of mechanical properties, it is useful to know the angle between directions.

1. Angle between directions can be determined by careful visualization and trigonometric calculations. In cubic systems, the angle can be determined from the relative simple calculations of a dot product of two vectors.
2. Linear density- Another quantity, calculation of linear density of atoms along a given direction.

Atoms are uniformly spaced along a given direction to determine the repeat distance γ between the attrition atoms. The linear density is simply the inverse γ-1.

Example- what is the angle between the [110] and [111] directions in the cubic system?

Linear density= number of atoms/ distance

Planar density = number of atoms / area

R = 0 .126 nm (atomic radius for iron)

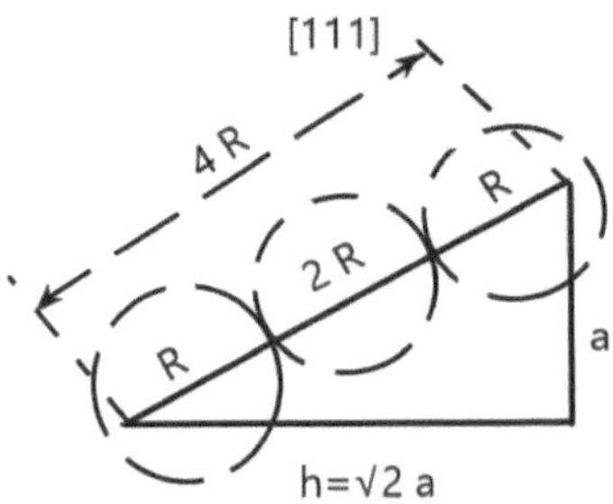

Fig 2.9

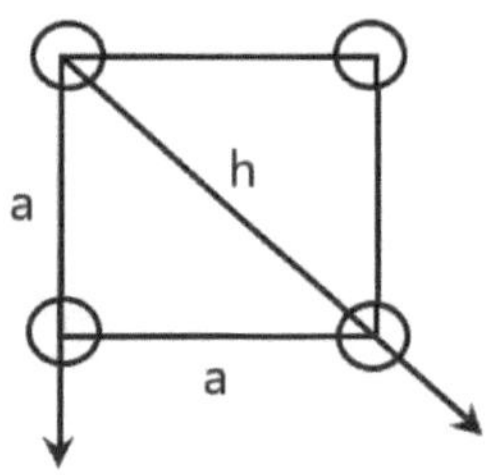

Fig 2.10

Planar density is used to find out slip plane. Slip occurred mostly dense plane greatest atomic packing (111) for more densely packed plane.

Crystal lattice are the infinite array of points arranged in space. These points can be joined together by drawing a straight line and by extending these lines in three direction. We notice that they appear to be a set of crystal plane.

The lattice points have different mechanical, electrical, or optical properties in different direction. This makes the study of crystal structure different. To overcome on this difficulty, we choose a set of crystal planes such that the properties of the crystal lattice remain unchanged in the direction of crystal plane.

In order to choose the specific crystal planes, Miller introduced a method known as Miller Indices. Crystals are anisotropic, so properties depend on the direction in which they are measured.

Close packed planes- It means more number of atoms per unit area which indicate structure really stronger giving highest density.
Slip planes with highest atomic density is preferred for deformation because the distance between atoms are small that dislocation movement due to lower applied stress is easier and higher atomic packing factor is an indication of easiness of deformation.

Packing crystal depends on sizes, charges, types, strength of bonds. FCC is having high packing factor in metals easily deformed, malleable, and less hard than ceramics. On the other hand, ceramics are harder because of ionic bond. Bond locks the atom particular ions so if you push it get cracked as to oppose the damage or dent.
Atomic planes can slide along each other, allowing for deformation because the bonds are metallic which are not very strong. Plastic deformation takes place by sleep or evening or both.
Slip- movement of dislocation in the atomic level
slip plane -plane has higher planar density of atoms
Slip direction -direction has highest linear density of atoms.
Combination of slip plants and slip direction is known as slip system.
Slip in BCC system occurs on the {110} {112} {123} with slip directions [111].
[6 slip planes of type {110} in each with two direction <111>
therefore 6x2=12+ there are 24 slip

So there are 48 slip systems in Bcc.
In FCC, slip systems= 12

In HCP, slip system =3

FCC and BCC have large number of slip systems so they are ductile while HCP has only few that is 3 slip systems so HCP is brittle.

BCC is having more slip system than FCC still FCC is ductile than BCC.
Because slip plane in FCC are more close pack than slip will plain in BCC. BCC{110} plane have higher atomic density but they are not much greater dens pattern as compared to other planes in BCC. <111> direction in BCC is closed pack as one <110> direction in FCC.
Packing in solids- [Close packed structures]
Packing of the spear in such a way that the atoms occupy maximum available space and there is minimum empty space. This type of packing is called close packing.
The principle of packing in solids-

In a crystal, particles (atoms, molecules, or ions) occupy lattice points in the crystal lattice. These particles may be of various shapes and hence mode of packing of these particles will change accordingly to according to their shape. Constituent particles are considered rigid in compressible sphere of equal size. The constraint particles try to get closely packed structure so that the maximum closely packed structure is attained. The maximum closely packed structure is such that the minimum empty space (void) is left. The closer the constraint particles lie, the greater is the stability of packed system. By this arrangement, a maximum possible density of the crystalline stability of the crystal is achieved.
Closed packing in one direction:

There is only one way of arranging spheres in one- dimensional close pack structures. Spheres can arrange in the row touching to each other, each sphere is in contact with two neighbour therefore coordination number is 2.

Close packing in two- dimension :

1. Square closed packing (A-A-A-A type)

Second row is in contact with the first one, we place more rows to obtain AAAA type.

2. Hexagonal closed packing

Second row may be placed above the first one in zig zag manner, such that spear fit in the gap of first stroke heck is a hexagonal clause packing have this free space and this packing is more efficient than square closed packing. Each sphere is in contact with six of its neighbour sphere, hence 2D hexagonal packing coordination number is equal to 6.

2.6 Closed packing in three dimensions-

A three- dimensional closed pack structure can be generated by stacking or placing layer of close practice spheres in two dimensions one over the other.

1. Three -dimensional closed packing from form a two-dimensional square layer

Arrangement of spears of both the layers are perfectly aligned, horizontally as well as vertically. Lattice has AAA….. type pattern. Coordination number for arrangement is equal to 6; volume occupied by particle is equal to 52% with a word space is equal to 48% that is simple cubic structure.

2. Three-dimensional closed packing from two dimensional hexagonal closed packed layer

Packing is done by placing a layer over the other.

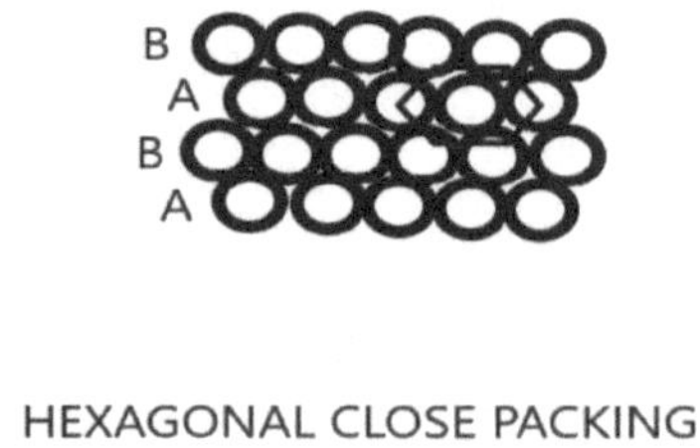

Fig 2.11

Fig 2.12

All the voids are equivalent in the first layer they have marked alternative as a and b. When the spear of 2nd layer is above the void of first layer a tetrahedral void is formed. This sort of arrangement of atoms is found in many metals.

3. Placing third layer over the second layer-

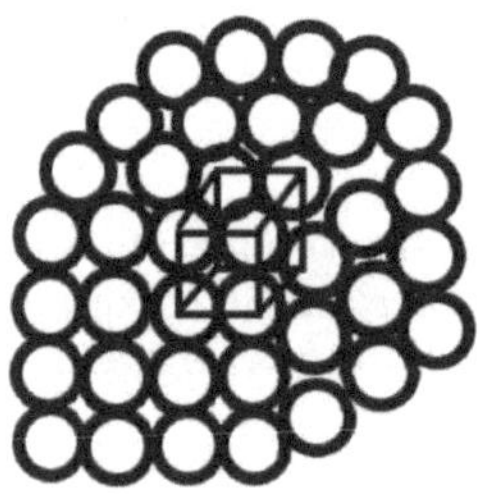

Fig 2.13

It is observed from the figure that the first layer will be considered as A and 2nd layer as B, then 3rd layer as C to the specific pattern of layers ABC ABC ABC pattern. Tetrahedral voids are formed where sphere of second layer is above void of first layer where 3rd layer is placed over second layer octahedral void the arrangement stands for CCP cubic close packing with packing efficiency 74%.

2.7 Line Defects/ Dislocations [one dimensional]

Any deviation from a perfectly. Of atoms along the line is called a line different or line imperfection called dislocations. Dislocations strongly affects the mechanical properties of materials.

Line defects are created during the solidification process, plastic deformation, vacancy, or atomic mismatch in solid solutions. These locations are of three types-

Edge dislocation
Screw dislocation
Mixed dislocation

Edge dislocation-

Extra half plane of atoms is present in crystal structure age or line. Vertical plane does not extend from the top of the bottom of the crystal. Because of incomplete plane of atom the atoms above the dislocation line are squeezed together and are in a state of compression whereas atoms below are pulled apart and experienced tensile stresses. Thus, region of compression and extension are associated with an age dislocation the bond length are compressed and stretched compared to normal value.

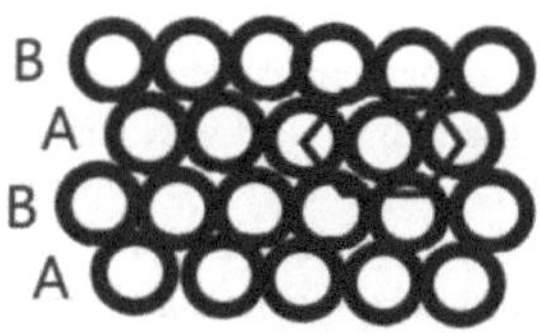

Fig 2.14

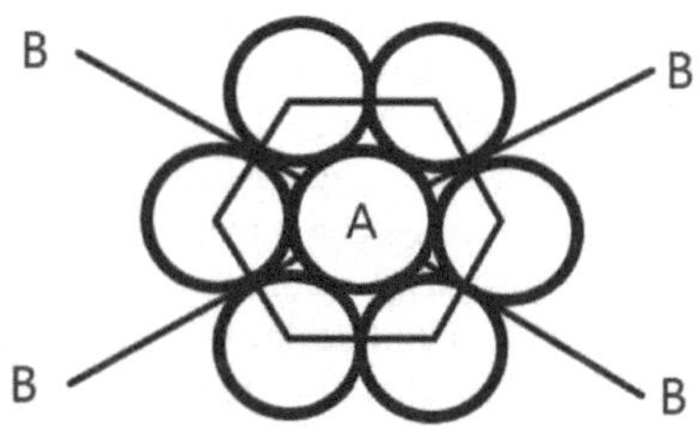

Fig 2.15

Bond lengths have been compressed two smaller than equilibrium value. Just below dislocation, attempts are pulled apart and are in the state of tension, the bond lengths have been stretched to above the normal value. This distorted configuration extends all along the edge into the crystal.

Potential energy increases from minimum value for the normal bond angle r0 to higher value on both sides that is for an increase or decrease in the bond length. Thus, there is extra energy due to distortion in the region immediately surrounding to the age of the incomplete plane.

As the region of maximum distortion is centered around the age of the incomplete plane, this distortion represents a line imperfection and is called Edge Dislocation, represented by $\perp$ or T depending on incomplete plane on top to bottom or bottom to top.

Depending on whether the incomplete plane starts from top or from bottom of crystal to configuration are referred as positive edge dislocation or negative edge dislocation.

Burgers circuit - represented by b-
Burger Circuit in perfect crystal

Burger vector (shortest distance) denoted by b that represent the magnitude and direction of the lattice distortion resulting from a dislocation in a crystal lattice, but since can be one way or other way.
So, Burger vector convention right or left hand side.

Screw dislocation-

Let the top part of the crystal be displaced by the atomic distance the left with respect to right part of crystal.
Screw dislocation at the boundary PQ or EF between the displaced and undisplaced part of the slip plane.
If the atomic plane of atoms follows spiral or helical path that is traced around the dislocation line, it is screwed dislocation. In screw dislocation, left front region of the crystal is shifted one atomic distance relative to right portion as shown in figure a whereas upper front region of the crystal is shifted one atomic distance relative to the bottom portion as shown in figure b.
Let t vector be defined such that it points from bottom to top, burger circuit is drawn in a clockwise sense as in figure a and right to left as in figure b in clockwise sense.
Burger vector b has same direction as positive vector, Burger vector b is parallel to tangent vector t, and therefore Burger vector is parallel to screw dislocation.
Burger Vector is a unit vector required to close loop around a dislocation. It indicates how much and in what direction the lattice above the slip plane appears to have been shifted with respect to the lattice below the slip plane.
Screw dislocation represented by clockwise direction or anti clockwise direction depending on whether the burger vector and tangent vector are parallel or anti parallel. This case refers to as positive or negative screw dislocation. If you go clockwise the circuit is taken clockwise you are going down this is what you will obtain if you curl the fingers of your right hand clockwise and the thumb is pointing down so that the right handed screw dislocation line.
Edge, screw and mixed dislocation-
Classification is based on the angular relation between b and t
Mixed dislocation- Practically pure edge or pure screw dislocation does not exist. They are in combined form called mixed dislocation. b perpendicular to T not perpendicular b parallel to T not parallel b and T are on the same angle mixed dislocation

Characteristic vector of a dislocation-

Every dislocation is characterized by these two vectors

One of them is called tangent vector or line vector and other is called Burger vector.

Distortion Energy or Dislocation Energy-

Dislocation has distortion energy due to compressive and tensile strength around the edge dislocation or shear strain around the screw dislocation. Strain is elastic strain. Elastic strain energy E per unit length of a dislocation of burger vector b is approximately given by-

Body centered cubic BCC-

The shortest distance between the lattice point is actually along the body diagonal, so half the body diagonal is become the shortest distance. So any of these vectors from any corner to nearby body center is burger vector. The entire body diagonal vector is <111>, we need half of that <111> so this is ½ <111>.

Simple cubic lattice-

<100>

family of 100 directions

for simple cubic lattice

Burger vector <100>

2.8 Surface Defects-

Surface defects are two- dimensional nature. Surface refers to region of distortion that lies about a surface having a thickness of a few atomic diameters. From thermodynamics point of view, surface defects are not stable and are present as metastable defects and under suitable conditions these defects can be removed. Many of surface imperfections removed by heatin, recovery. Defects occur due to number of bond broken at the surface, imperfection due to different crystal structure, crystallographic orientations, slow changes in crystal during solidification. It may arises from the clustering of line defects bracket due to age or screw dislocations.

Singular surface has low surface energy and usually have low index plane such as (100), (110) or (111). Vicinal surfaces have orientations near those of singular surfaces and separated by monoatomic steps. Orientation of unit cells of one grain changes to other that is at this surface there is a mismatch of unit cells with unlike orientations and tends to higher interfacial energy depends on grain boundary angles.

Classification of surface or interfaces defects

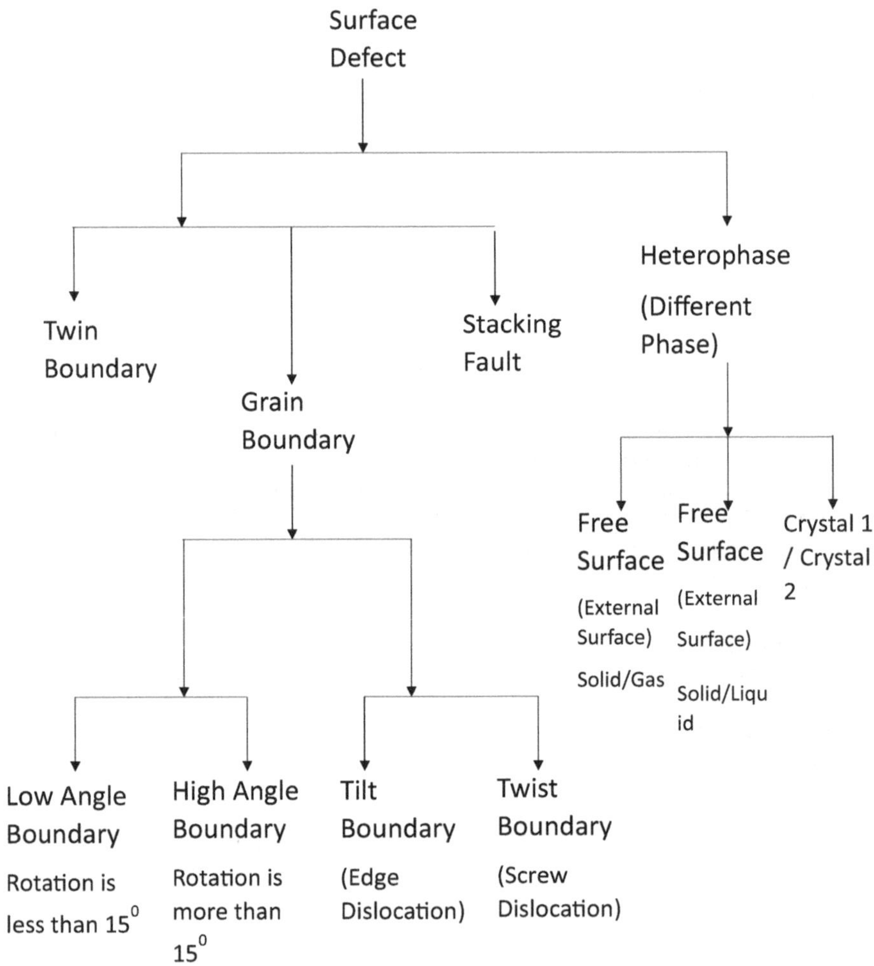

Grain boundary

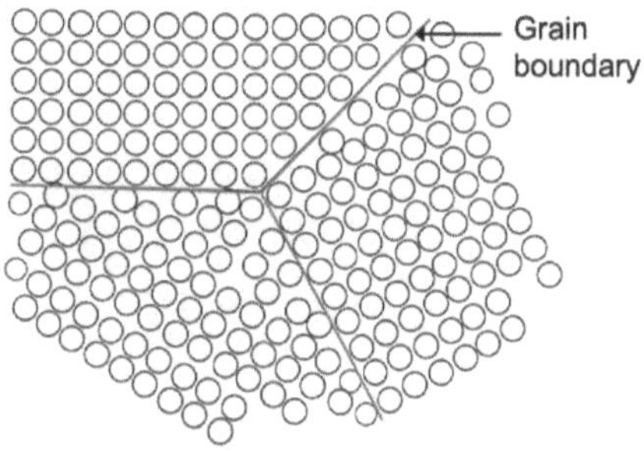

Fig. 2.16

External surface or free surface-

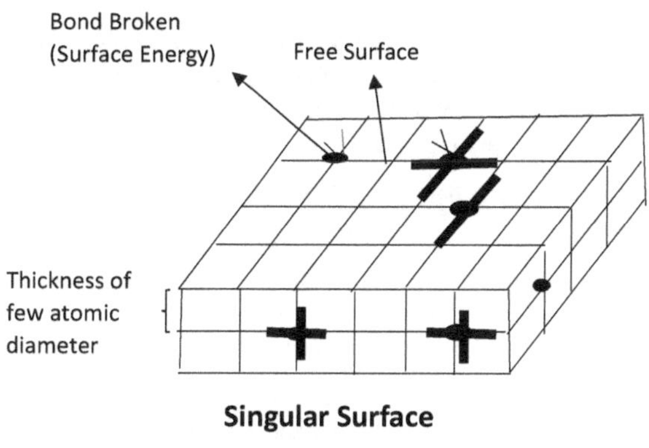

Fig. 2.17

The environment of an atom at a surface differs from that of an atom in the bulk, mainly the number of neighbors that is coordination at surface is less. The unsaturated bonds of surface atoms give rise to surface energy.

If any crystal bounded by surface, its inside crystal may or may not perfect but crystal surface will contain imperfection. On the surface as there is an external boundary, as that external boundary breaks the crystal lattice (due to broken bond at the surface) is the defect.

Every external surface is related to breaking of a bond therefore in two dimensional models inside atom is bonded with four neighbor atoms but on the surface it is bonded with only 2 atoms as other two atom bonds are broken shown by dotted lines.

Breaking of bond requires energy this energy is extra given to the surface is called surface energy. Surface energy quantifies the disruption of intermolecular bond that occurs in a surface is created. Also called left surface free energy. In simple language, surface energy can be defined as the work per unit area done by the force that creates the new surface.

Grain boundary

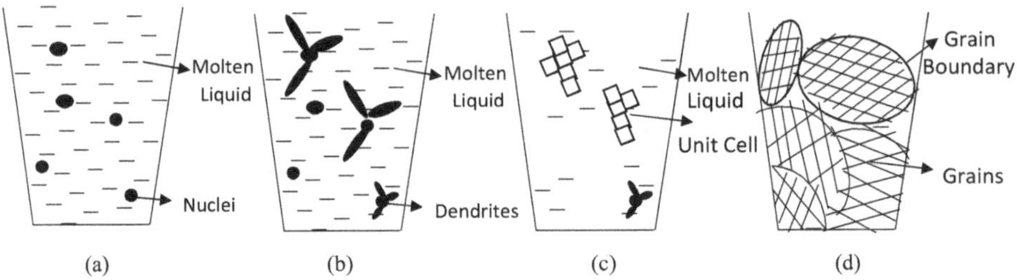

Fig. 2.18

1.Crystalline solids are usually made of number of grains separated by grain boundaries. Boundaries are of several atom distance wide and orientation of grains are mismatched. When this misalignment is small on other side with respect to first generally orientation angle is up to 10 degrees it is called low angle grain boundary or tilt boundary.

2.Tilt boundary causes a constant angle of tilt between lattice planes of the same type. Misalignment axis is parallel to plane boundary. Misalignment composed of edge dislocation.

3.When the misalignment of grain atom is larger or larger orientation more than 15 degree it is called as high angle boundary degree of disorientation is of large range greater than 15 degrees, in which this orientation axis perpendicular to boundary plane, therefore it is twist boundary.

4.Rotation axis of plane may be parallel or perpendicular according to tilt or twist boundary respectively. Tilt less than 10degree angle of rotation and twist angle more than 50 angle of rotation.

4.Crystal is separated by horizontal axis lying in a particular plane perpendicular to the tilt boundary. Tilt boundary means boundary plane such that the grains on either side are rotated with respect to each other such that the rotational axis lies in the boundary plane, orientation is shown in figure, so rotation axis is perpendicular to the plane of paper. This location represents set of parallel planes. If rotational axis vertical is parallel to the plane of paper, that is set of a particular plane is termed as screw dislocation.

Twin Boundaries-

Twin boundaries are type of planar surface defects. Atomic arrangement on one side of a twin boundary is mirror reflection of the arrangement on the other side. Shear force is applied atom traveled and creates mirror image. Twin boundaries occur in pairs. The main two reasons of twin boundaries are

1. phase transformation where cooling from a high symmetry phase at high temperature

2. mechanical tuning due to stresses that is deformation

Boundary in a crystal such that crystal on either side and mirror images of each other is called as twin boundary.

Edge and screw dislocation occurs due to slip planes.

Twin boundary occurs due to shear stress.

Twin is bend at lower temperature and mainly hexagonal closed packing twin is crying.

Twin boundary in CCP structure-

Fig.2.19

Creating a new sequence which is mirror like sequence of the crystal of other side of mirror plane. This changed the orientation, two orientation are related to each other by mirror reflection.

CCP[PERFECT]	CCP WITH A TWIN BOUNDARY
C	C
B	B
A	A
C	C
B	B
A	A Mirror Plane

C	Reflected		B	[Twin Plane]
B	Different Orientation	C		
A		A		
C		B		
B		C		
A		A		

Stacking faults-

There is no change in orientation of crystal across the stacking fault. Crystal on one side of the fault is shifted by a non lattice translation with respect to the crystal on the other side.

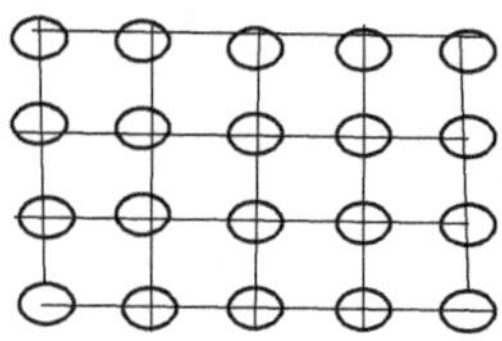

Fig 2.19

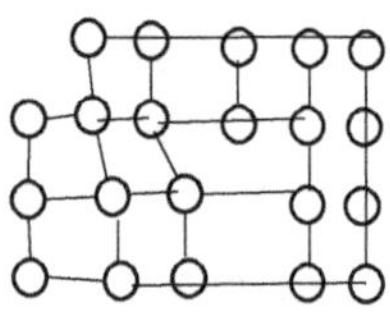

No hanging plane

Fig. 2.20

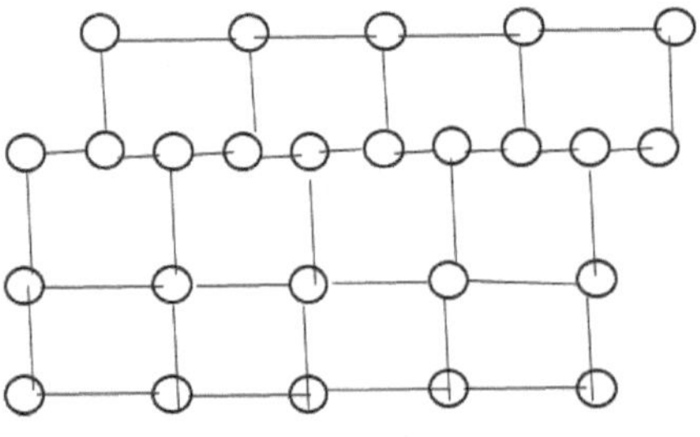

(Hanging plane)

Fig. 2.21

Grain boundaries rotational boundary whereas stacking fault is the translational bond .

Stacking sequence relation to closed packed structure-

Stacking sequence in a cubic close packed structure CCP-

Perfect sequence in close packed structure arrangement of number of planes

These type of defect naturally occurs in crystal when the crystal is growing by addition of new planes, affect energy which create defects known as stacking defects.

If plane is missing for example c plane is missing it is intrinsic stacking defect fault and if extra plane not the part of the bottom or top plane is added then it is extensive stacking fault.

CHAPTER 3

Imperfections in solids

3.1 Imperfections in Solids:

Vacancies, equilibrium. The concentration of vacancies, interstitial and Substitutional impurities in Solids, dislocations, types and characteristics of dislocations, Interfacial defects, and stacking faults.

Crystal Imperfections- Non-Crystalline structures have no regularity or periodicity in the atomic arrangement.

Ideal Crystal: An ideal crystal is a crystal with a perfectly regular lattice containing no impurities, imperfections, or other defects. A perfect crystal is a crystal that contains no points, lines, or planar defects.

HYPOTHETICAL CONCEPT OF A PERFECT CRYSTAL IS IMPORTANT (3rd Law of Thermodynamics)

[The third law of thermodynamics states that the entropy of a system approaches a constant value as the temperature approaches absolute zero.]

It is impossible for any process, no matter how idealized, to reduce the entropy of a System to its absolute zero Value. Various thermodynamic processes create imperfections.

Real crystals- Real crystals contain lattice imperfection (do not fulfill the perfect repetition of unit cell in three dimensions) which affects those properties sensitive to structure (some of the atomic bonds are broken). If volume-wise imperfection is 0.01% of total volume, insensitive properties such as melting point, specific heat, and density do not affect it. The density of the crystal would be lower by the same amount (i.e. 0.01%) which is small enough to be ignored.

However, crystalline materials that deal with engineering practice, its properties that are sensitive to structure, such as tensile strength, hardness, electrical resistivity, and thermal conductivity at lower temperatures. For example, 1/1000 (few parts/millions of aluminium - Al impurity may change the character of a silicon sim conductor. If impurity gets in unintentionally or accidentally, it could destroy the semiconductor devices such as transistors or rectifiers of junction together. Similarly, line imperfections known as dislocations, decrease the mechanical strength of crystals significantly.

The defects in atomic structure and arrangement are called imperfections. These imperfections result from the deformation of solids, rapid cooling from high temperatures, or striking of solids. Heat treatment of metals makes it possible to refine the properties.

Crystalline imperfections can be classified on the basis of their geometry as follows-

1] Point Imperfection- Zero-dimension defect

2] Line imperfection- One- dimension defect

3] Surface imperfection- Two- dimension defect

4] Volume imperfection- Three- dimension defect

Volume imperfections can be foreign particle inclusions, large voids or pores or non-crystalline regions.

Units

Sr No	Quantity	Symbol	Unit
1	Concentration of point imperfection	n/N	-
2	Enthalpy of formation of point imperfection	ΔHf	KJ/mol
3	Dislocation density	ρ	m/m3
4	Dislocation energy	E	J/m
5	Shear Modulus	μ	N/m2
6	Energy of surface imperfection	ϒ	J/m2
7	Surface Tension	π	N/m2

3.2 Defects

Point Imperfection are zero-dimensional imperfection. As the name implies, they are imperfect point line regions in the crystal. Different kind of point imperfections are described as-

[A] VACANCY:

Vacancy refers to an atomic site from where the atom is missing as shown in figure. Vacancy is a defect occurred due to missing atom in an atomic structure plane. Vacancies results during solidification and due to atomic vibrations or thermal disorder. This causes displacement of atoms from their usual lattice location, when an atom leaves its site and gets placed in the interstitial sites of atomic structure, a self-interstitial atom and defects occurs.

A substitutional impurity (or solute) is a point imperfection. It refers to a foreign atom that substitutes for or replaces a parent atom in the crystals as shown in figure b. Aluminum and Phosphorous doped in Silicon are substitutional impurities in the crystal.

An interstitial impurity is also a point imperfection. It is a small sized atom occupying the void space in the parent crystal, without dislodging any of the parent atoms from their sites, as shown in figure C. An atom can enter the interstitial void space only when the atom enters the interstitial void space only when it is substantially smaller than the parent atom.

In closed packed structure, the largest atom that can fit the octahedral and tetrahedral voids have radius of 0.414 rand 0.225r respectively, where r is the radius of the parent atom. Example- Carbon is an interstitial solute in Iron.

Enthalpy of formation of vacancies:

The presence of a point imperfection introduces distortion in the crystals. If the imperfection is a vacancy, missing atom forms no bond or bond breaks and as in case of impurity atom, result of size difference will be atoms due to impurity atom, created elastic strain in the region of crystal immediately surrounding the impurity atom.

The elastic strains are present or created irrespective of whether the impurities atoms are larger or smaller than the parent atom. A larger impurity atom introduces compressive stress and corresponding strains around it, while a smaller atom creates a tensile stress-strain field.

As these factors tends to increase the enthalpy (or potential energy) of the crystal. The work required to be done for creating a point imperfection is called the enthalpy of formation (ΔHf) of point imperfection. It is expressed in kJ/mol or eV/point imperfection.

Enthalpy of formation of vacancies in some crystals-

Crystal	Al	Ag	Cu	Ni
ΔHf	68	106	120	168
eV/vacancy	0.7	1.1	1.24	1.74

1 eV= 1.602×10^{-9} J

Entropy of a Crystal

Gibb's Free Energy [Gibb's energy is used as a criterion of stability]

Gibb's Free Energy in terms of enthalpy (H) and entropy (S)

$G = H - TS$

Gibb's Free Energy represents the available part of the energy which can be converted into work.

Suppose n point imperfection is introduced in one mole of crystal, the change in Gibb's free energy ΔG is-

$\Delta G = G_{final} - G_{initial}$

$= G_{defect} - G_{perfect}$

$= \Delta H - T\Delta S$

Enthalpy of formation and entropy of crystal (configurational entropy) are used to find Gibb's energy. Most stable state of the material is having minimum Gibb's energy.

For equilibrium concentration; $d\Delta G/dn = 0$

$n/N = \exp^{(-\Delta Hf/RT)}$

Possible arrangement of Sixteen spheres, eight of which are white and eight are black.

It is easy to calculate total number W of such different configuration.

Suppose, total number of sites as N, number of white spheres as n, number of black spheres will be (N-n).

Number of configurations, $W = \frac{n!}{(N-n)!\,n!}$

(eg if N=16, n=8 then W= 12870)

One mole of a solid contains more than 10^{23} atoms.

If we mix two different kind of atoms randomly in a solid, large number of different configuration and large amount of configuration energy will exist.

Suppose a mole of solid with atomic site N contains n atoms of B and (N-n) atoms of A are mixed randomly, configuration entropy is zero before mixing. Increase in entropy due to mixing is given by

ΔS = S final - S initial = S defect- S perfect

=> $\Delta S = K \ln W$, where K= Boltzmann Constant = 1.380 x 10^-23 J/K

W= n!/(N-n)! n!

ΔS = K ln W= n!/(N-n)! n!

ΔS = K ln N!/(N-n)!n! ---------- A

ln n! = n ln(n) - n ---------- B

Subs. B in A:

ΔS = K[(N ln(N) - N) - ln[(N-n)!n!]]

ΔS = K[N ln (N) - (N-n)ln(N-n)-n ln(n)]------ C

The equilibrium state of the crystal will corrsponds to the minimum in its free energy (Gibbs free energy)

We get it when $d\Delta G/dn=0$

As we know, $\Delta G = \Delta H - T\Delta S$ ---------- D

Subs. C in D

ΔG = ΔH - KT[N ln (N) - (N-n)ln(N-n)-n ln(n)]

n/N=exp[-Hf/RT]

Equilibrium Concentration = n/N=exp[-ΔHf/RT]

Q) Find the equilibrium Concentration of vacancies in aluminium and nickel at 0 K,300 K and 900 K. ΔHf for Al=68KJ/mol and ΔHf for Ni=168KJ/mol.

n/N=exp[-ΔHf/RT]

Final values:

Temperature	0 K	300 K	900 K
Al	0	1.45 x 10^-12	1.12 x 10^-14

Ni	0	5.59 x 10^-30	1.78 x 10^-10

Number of vacancies is the function of energy vacancies have been seen to be thermodynamically stable in crystalline solids above 0° Kelvin and their number increases exponentially with the increase of temperature. For this concept, free energy is used.

In perfect crystal, all lattice sites are occupied by atoms so that no vacancies are present. If E_f is the energy required to create a vacancy for removing an atom from the lattice site, the total energy increased due to formation of N such defects are nE_f.

Let there be N atoms on the lattice sites, E_f is given in kJ/mole or eV/defect.

Equilibrium concentration of vacancies as a function of temperature for aluminum-

The equilibrium vacancy concentration at different temperature for different configuration entropy is as shown in diagram which shows that vacancy concentration increases rapidly as the temperature increases. Distance between internal atoms increases, mutual attraction is reduced, plastic deformation occurs. Vacancy concentration affect the material strength.

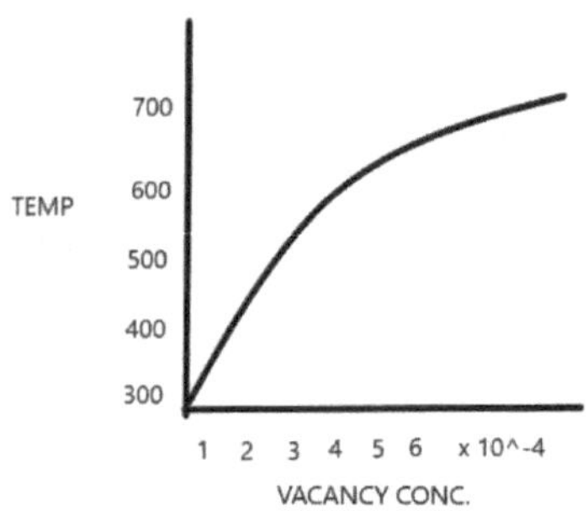

Fig. 3.1

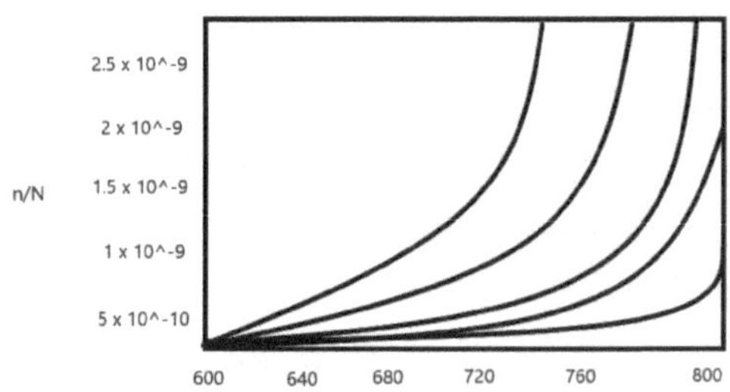

Fig. 3.2

CHAPTER 4

The anatomy of Materials

Structure of Materials and strength of materials, Yield strength, tensile strength, and ductility of materials: stress, strain behavior of metals, ceramics, and polymers, tensile test, plastic deformation, necking, creep behavior, and fatigue.

4.1 Introduction-

When a solid object is under stress, it undergoes deformation. Strain is a measurement that gives the change in length of an object divided by the original length. Strain is usually given by the symbol ε. If the sample of material is subjected to different levels of stress, corresponding strains are measured and the graph of stress vs strain curve will be the characteristic curve for a given material.

The graph of stress-strain curve for malleable material such as mild steel-

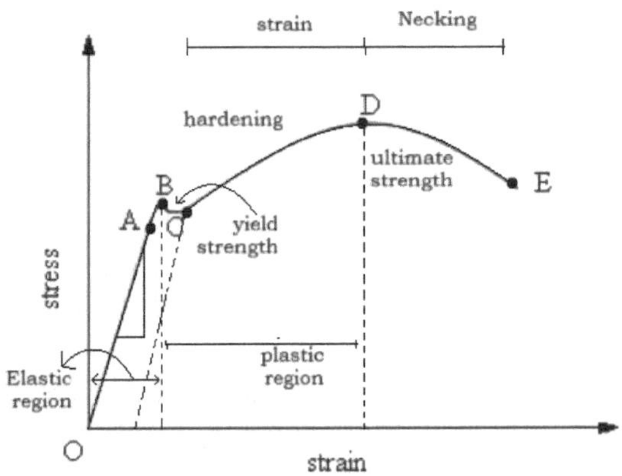

Fig. 4.1

When the stress on a material is slowly increased, the corresponding strain increases in proportion in the beginning. If the force causing stress on the material is removed then the material would return to its original shape. When a material can do this, the material is elastic.

The stress on the material keeps increasing, then the material would eventually reach a point when the material becomes so deformed that, even when the deforming forces are removed, the material is unstable to return to its original shape. The stress at which a material stops behaving elastically is called yield strength.

When the material is unstable it returns to its original shape, but is plastic.

Proof stress is very similar to yield stress for materials that do not exhibit yield points. Point on the lines similar to the way yield stress but with a specific amount of plastic deformation.

Proof stress value on the strain axis usually (0.2%) parallel line is shown to the linear part of the stress-strength curve for ductile materials such as aluminum, copper, and silver.

Brittle materials such as concrete do not have well-defined yield products therefore ultimate strength or breaking strength is the same as glass which is brittle material do not show any plastic deformation but fails when the deformation is elastic.

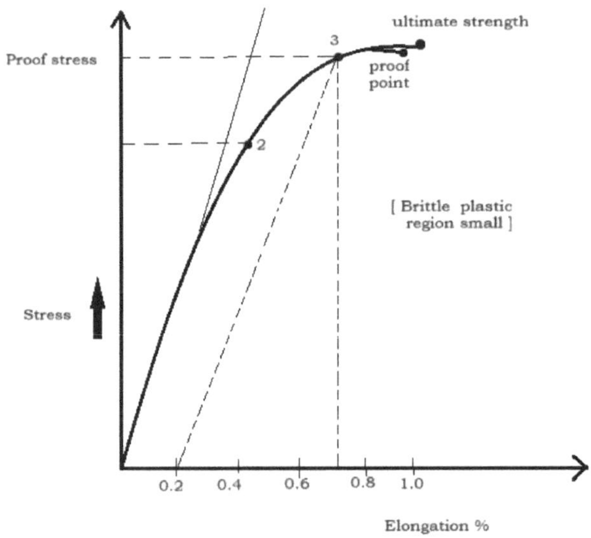

Fig. 4.2

Yield strength is the stress of material can withstand without permanent deformation. Yield point at which material transforms from elastic to plastic which will be the end of elastic behavior and beginning of plastic behavior.

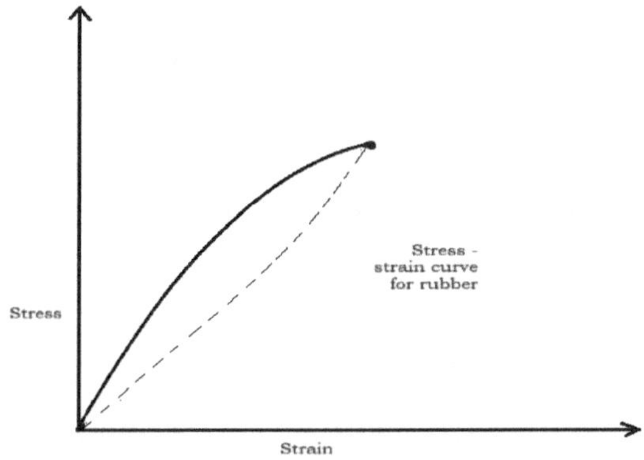

Fig. 4.3

Knowledge of materials yield strength is very important when designing components since it usually represents the upper limit of the load that can be applied without producing permanent deformation. It is useful for controlling material production techniques such as forging, rolling, and pressing.

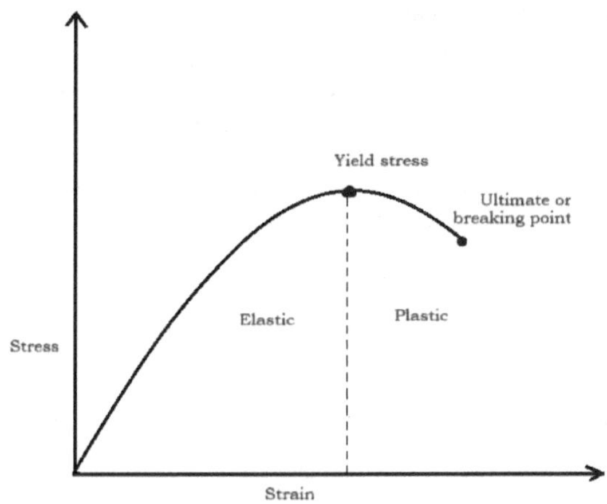

Stress - Strain curve plastic

Fig. 4.4

Material	Yield Strength (MPa)	Ultimate Strength(MPa)	Density (g/cm3)
Mild Steel	248	841	7.58

Human Skin	15	20	2.2
Stainless Steel	520	860	8.19
Cast Iron	130	200	7.2
Copper	70	220	8.92
Glass		33	2.53
Concrete		3	2.57
Human Hair		380	
Diamond	1600	2800	3.5

4.2 Tensile Strength-

Tensile strength is the maximum load that a material can support without fracture when being stretched in the plastic region (or it is the maximum strain that a material can withstand while being stretched or pulled before breaking).

The tensile strength also known as the ultimate tensile strength, is load at failure divided by the original cross-sectional area. Ultimate tensile strength (UTS) =

As its name implies, tensile strength is the material resistance to tension that is caused by mechanical loads applied to the material. The ability to resist breaking under tensile stress is one of the most important and widely measured properties of materials used for structural applications. For example, if a metal rod one square inch in cross-section area can withstand a pulling force of 1000 pounds but breaks if more forces are applied, the metal has a tensile strength of 1000 pounds per square inch. The tensile strength for steel is 400 MPA.

Three types of tensile strengths are-

Yield strength- the stress a material can withstand without permanent deformation

Ultimate strength- the maximum stress of material can with strength

Breaking strength –the stress coordinate on the stress strain curve at the point of failure.

Performing a tensile strength test for metal will determine how much an alloy will elongate before reaching its ultimate tensile strength. How much load a particular piece of metal can accommodate before it loses structural integrity.

Therefore, understanding the tensile strength of our material is important for construction and personal safety. Tensile strength is important parameter for materials that are used in structures and mechanical devices. Specified for building materials such as alloys, composite materials, ceramics, plastics, and wood.

4.3 Ductility of materials-

Ductility is defined as the ability of a material to deform plastically before factory. It is the property of metal by which it can be drawn into wires. Two measures of ductility are elongation and reduction of area.

Ductility is the ability of a material to sustain a large permanent deformation under a tensile load up to the point of fracture. The ability of a material to have its shape changed as by being drawn out into wire or thread without losing strength or break. Examples- most metals are a good example of ductile materials including gold, silver, copper.

High degree of ductility occurs due to metallic bonds, which are found predominantly in metals. Leads to common perception that metals are ductile in general. In metallic bonds, valence shell electrons are delocalized and shared between many atoms. Non -metals do not show ductility due to absence of metallic bonds which allow one electron to slip over another without being subjected to strong repulsive forces that would cause nonmetals to shatter. Metallic bonds have more slip systems, becomes almost equally easy to activate by external stresses.

Measurement of ductility

1. Elongation
   % elongation= change in length/ original length x 100
2. Reduction in area
   % Reduction in area= change in cross sectional area/ original cross sectional area x 100

   Elongation is measured by the change in gauze length while reduction in areas measured by the difference between the original cross section and final cross section areas.

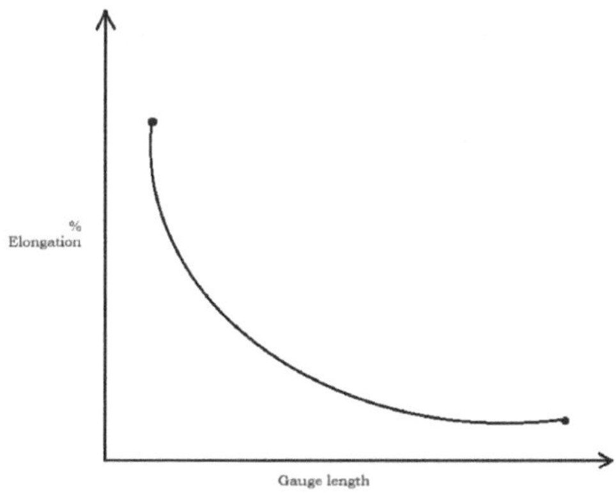

Fig. 4.5

Problem- A mild steel bar is subjected to tensile test. Boss length was atm, initial diameter was 12M the load at yield point was 3750 kg the maximum load observed

5500 kg and the breaking load was 3900 kg. Final length at breaking point was 100 mm. The diameter of the sample was 8mm at the fracture surface. Calculate the following-

yield stress, UTS ultimate tensile stress and breaking stress in mpa, ductility and percent reduction in area.

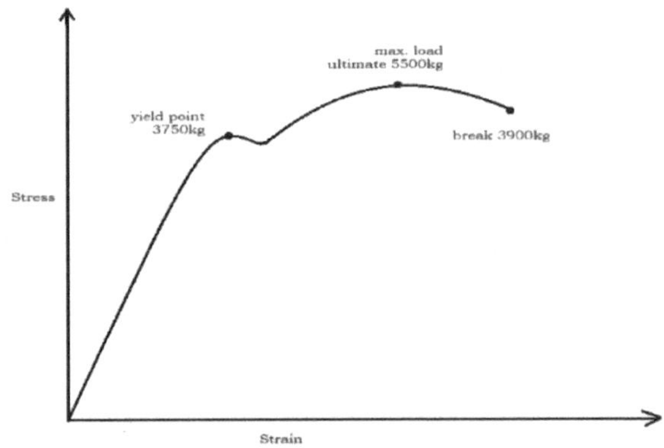

Solution: L_o= 80mm, d_o =12mm, A_o= $\pi/4$ x $d_o{}^2$ =113.14 mm^2 = 113.14 x 10^{-6} m^2

Now, load at

Yield point = 3750 x 9.81 = 36,788 N

Ultimate tensile strength = 5500 x 9.81 = 53955 N

Breaking = 3900 x 9.81 = 38259 N

Hence , σ_{yield} = 36788 / 113.14 x 10^{-6} = 325.15 M Pa

σ_{UTS} = 53955 / 113.14 x 10^{-6} = 476.89 M Pa

$\sigma_{breaking}$ = 38259 / 113.14 x 10^{-6} = 338.16 M Pa

Ductility : % Elongation = change in length / Original length = $\frac{100 - 80}{80}$ x 100 = 25%

% Reduction in area = $\frac{A_o - A_f}{A_o}$ x 100

= $\frac{113.14 - 50.28}{113.14}$ x 100

= 0.556 x 100

= 55.6%

4.4 Testing-

Destructive testing -in destructive testing sample get destroyed during the test

Non-destructive testing- in non- destructive testing sample remains same after the test as it was before the test.

Mechanical testing of metals is carried out for the evaluation of their mechanical properties. The composition, processing and heat treatment influence the properties of

metals. Important mechanical properties include strength, stiffness, elasticity, plasticity, ductility, brittleness, hardness etc.
Mechanical testing is required for the following purposes

1. study of failure analysis
2. research and development
3. testing after heat treatment and processing.

Important mechanical test includes-

- Tensile test
- Compression test
- Shear test
- Torsion test
- Hardness test
- Impact test
- Creep test and
- Fatigue test.
- Tensile Test-

This test is frequently used to evaluate certain properties of metals. This test has the great importance in the design of material for structures and other engineering applications.

4.5 Principle of tensile test- if a metallic specimen is subjected to a gradual increasing uni axial tensile load, it gets plastically deformed and finally fails or breaks. During its plastic deformation, changes occur in cross sectional area and length. Strength and ductility are the main properties which are usually judged in this test.

Procedure- in tensile testing, the metallic specimen is gripped at the opposite end and is pulled apart. Specimen elongates in the direction of applied load. A standard specimen is prepared. The specimen has a circular cross section. The specimen may be having rectangular or square cross section, if it is obtained from a plate. At the end of specimen, the failure may occur due to stresses introduced by the gripping device. End of the specimen should be suitable for proper gripping.

Tensile test has a useful scope in determining the following properties of a metal-

Proportional and 1]Elastic limit

2]Ultimate yield strength

3]Maximum stress

4]Breaking stress

percent elongation and reduction of area etc.

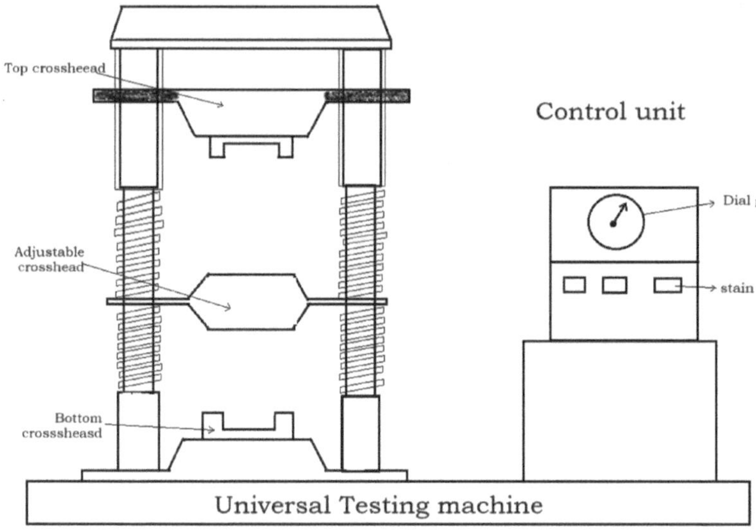

Fig.4.6

With the increase in load, there is a corresponding increase in length between the two extremities of the gage length. Plot a graph with the loads or stresses along the vertical axis and corresponding elongation or strain along horizontal axis, we will obtain a curve known as stress- strain diagram.

Proportional limit- OA is a straight line which represents stress is proportional to strain. Yield point- materials stressed beyond point B, plastic stage will be reached.

Proof Stress -in most of the material, they do not have a clear- cut yield point and for such material, proof stress is calculated.

Ultimate stress - stress goes on increasing till the point D is reached, attain its maximum value, neck forms.

Breaking stress- a specimen breaks away at Point E.

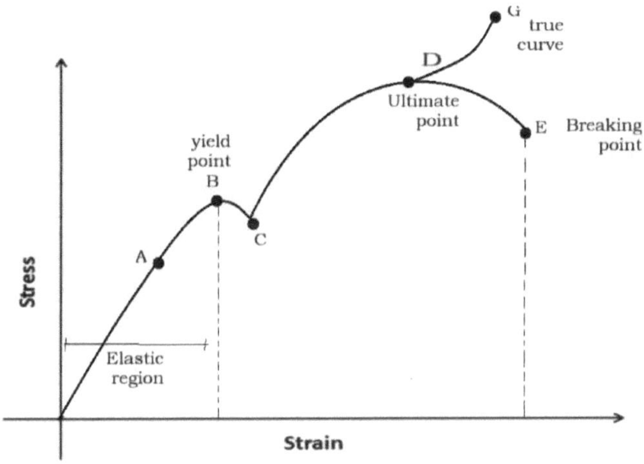

Fig 4.7

4.6 Yield strength-

Stress at which a material stops behaving elasticity called yield strength. Men stresses exceed the yield point, still will not come back represent a per limit for designing component that is strain hardening.

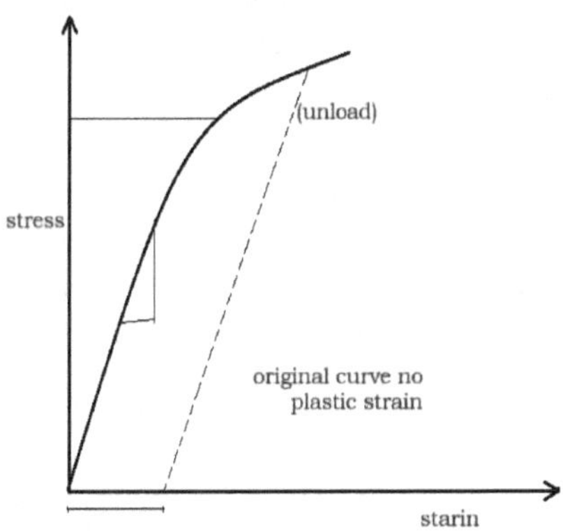

Fig. 4.8

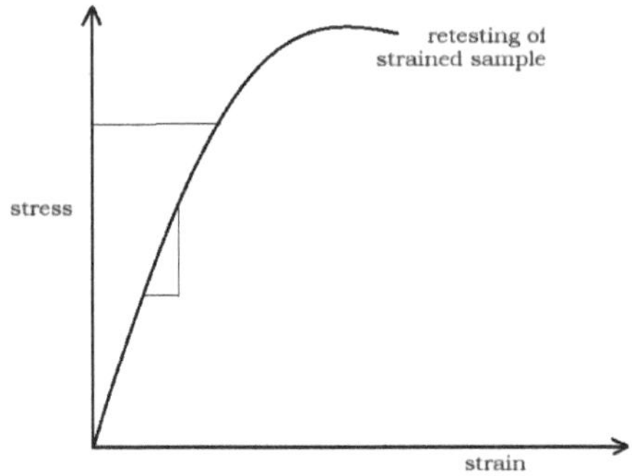

Fig. 4.9

Sample deformed elastically further curve become nonlinear and plastic deformation began, continue plastic deformation, deformed sample or material band yield stress little bit then stop the test and upload the material before fracture unloading curve is parallel is the plastic strain remain in the material.

Repeat the tensile test for deformed material, loading the strain sample or retesting the strain or deformed sample. Its yield stress continues further beyond the original yield stress, it goes up to unloading that point is new yield stress. Deformation where test is performed on deformed sample have higher yield stress. This term is strain hardening.

Yield strength affected by the following factors

strength hardening and

temperature of metal and microstructure.

Microstructure means size of the grains, small grain size has higher heels. It can be controlled by recrystallization and other techniques. At room temperature, hardness, yield strength, tensile strength, fatigue strength, and impact strength- all increases with decrease in grain sizes. Higher the cooling rate, smaller will be the grain size during solidification.

4.7 Hall Petch equation-

$\sigma_y = \sigma_o + Kd^{-1/2}$

Where, σ_y= Yield strength of polycrystalline

σ_o=Yield strength of single crystal

K = Constant

d= grain size

Graph shows with the decrease in grain size , yield strength increases.

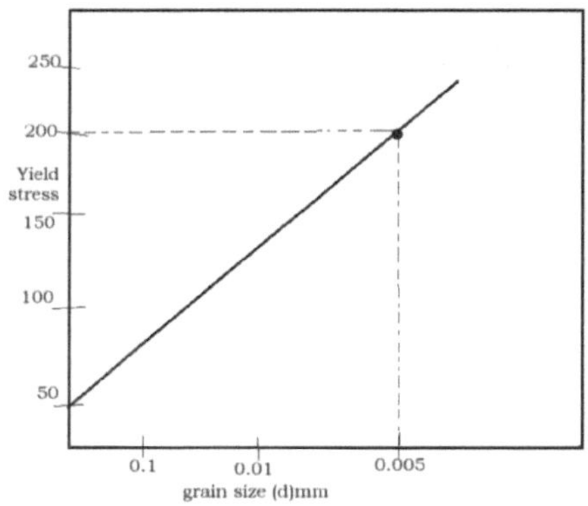

Fig. 4.10

4.8 Stress- strain curve-

The results of tensile tests are expressed in the graphical form which is called as stress -strain curve. Stress is defined as the intensity of internally distributed forces that resist a change in shape of the metal; whereas strain is defined as a change per unit length in a linear dimension of the metal. The results of tensile strain are also plotted as applied load kgs versus extensions mm. These are called as load extension diagram.

Stress strain values are derived by load and extension values as follows

Stress = applied load/ original cross sectional area

strain = change in length/ original length

True stress strength curve- true stress strength calculation the changing cross sectional area is considered.

True stress = load/ instantaneous cross sectional area at that load.

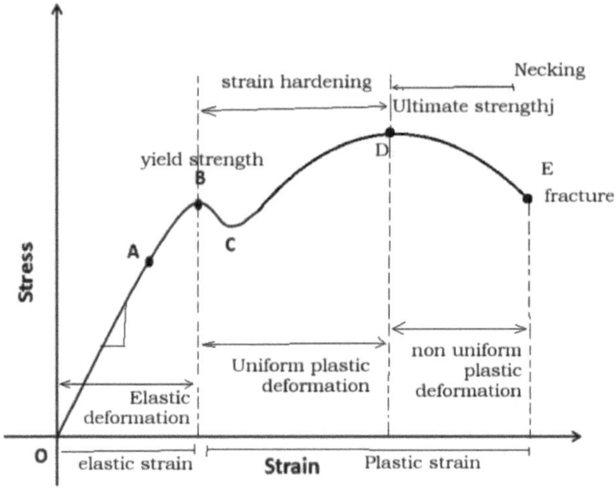

Fig. 4.11

<u>Relation between engineering and true stress strain</u>

Engineering strain(e) = ΔL\ Lo ———————— (1)

Total length L= Lo + ΔL

∴ ΔL = L - Lo ———————— (2)

∴ e = L – Lo \ Lo ———————— (3)

e = [L\Lo -1]

L \ Lo = e + 1 ———————— (4)

$$\epsilon_{true} = \int_{L^0}^{L} \frac{dL}{L} = \ln L/Lo \qquad \text{———————— (5)}$$

Substituting 4 in eq^n 5

$\epsilon_{true} = \ln(e+1)$ ———————— [A]

Similarly, $\sigma_T = P/A_i$ and $\sigma_E = P/A_o$

Taking ratio

$\sigma_{T/}\,\sigma_E = Ao/A_i = L/Lo$

$\sigma_T / \sigma_E = L/L_O = e + 1$

$\therefore \quad \epsilon_T = \ln (e + 1)$

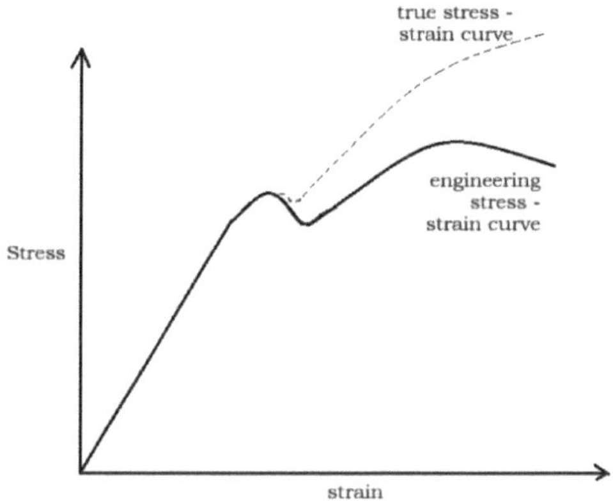

Fig.4.12

(A typical true stress strain curve for mild steel superimposed on engineering strain stress curve.)

Problem to solve - A tensile specimen of aluminum alloy rod having 5cm length and 1.5 cm diameter is subjected to maximum load of 37000 N. If diameter at fracture was 1.2 cm, calculate the following-

Tensile strength

True stress

True strain.

Some typical stress- strain curves for various metals, alloys and non – metals are shown in figure.

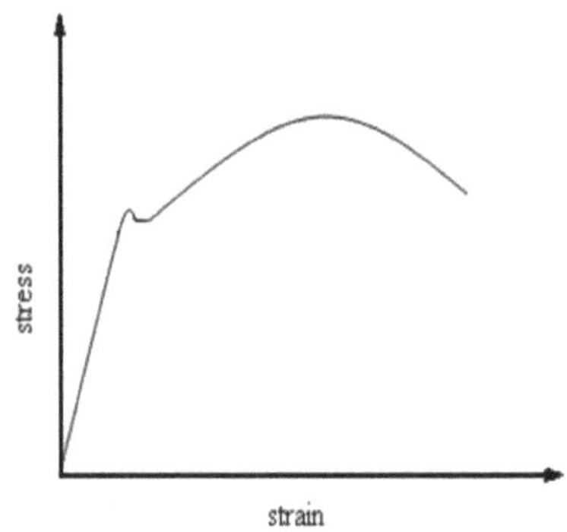

(Ductile metal with yield point)

Fig. 4.13

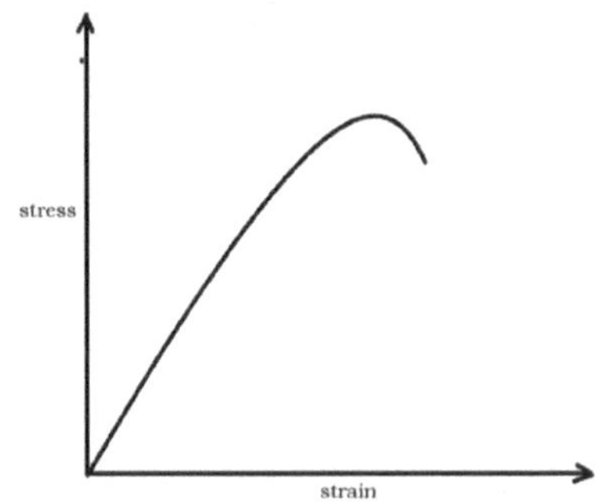

(Ductile metal without yield point)

Fig. 4.14

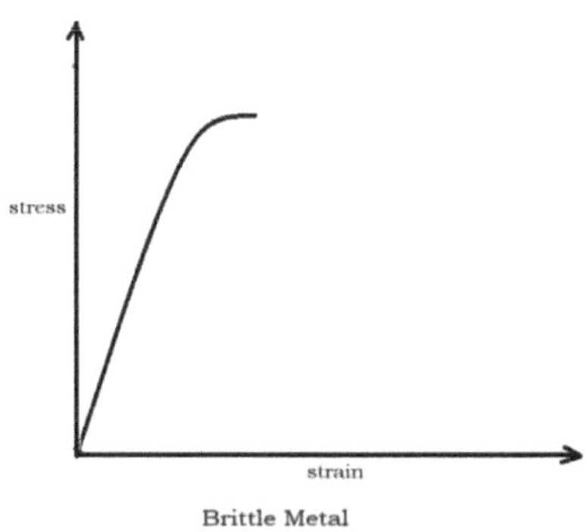

Fig.4.15

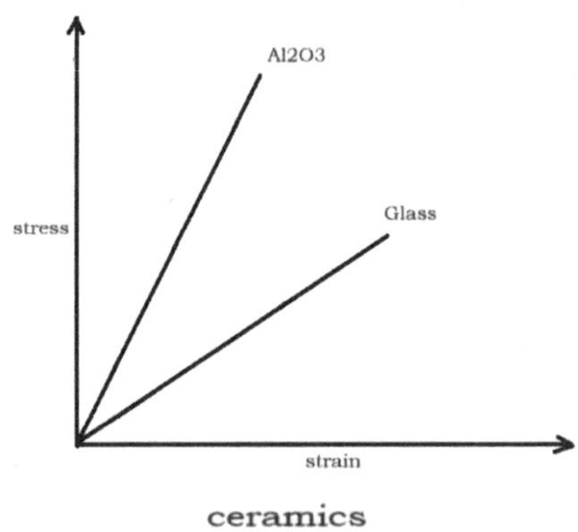

Fig. 4.16

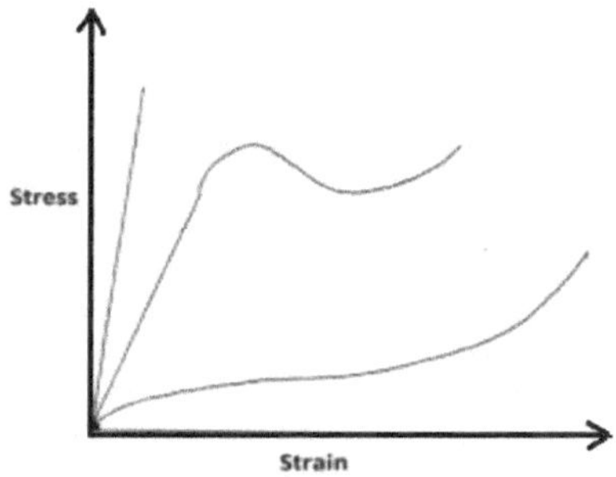

Polymer

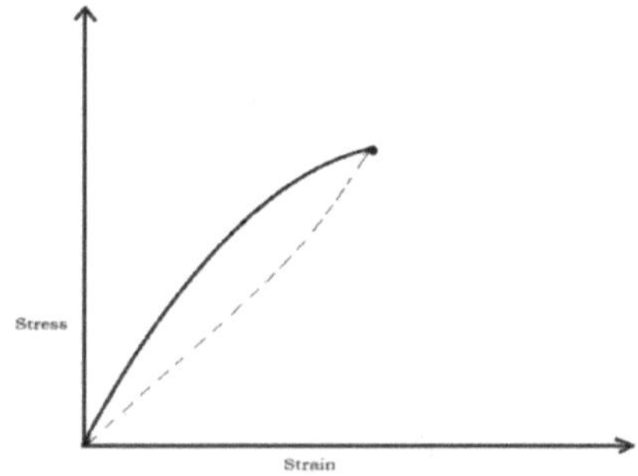

Rubber

Fig. 4.17

Fig. 4.18

(Various stress – strain curve)

4.9 Necking-

Necking is observed in the specimens during tensile testing. Usually, it occurs in the central portion of the specimen. When the metal specimen is pulled apart by the application of load, it elongates. The length of the specimen increases, while it overall cross- sectional area decreases. After certain maximum load, the decrease in local cross sectional area is observed. This is called as necking. The failure of metal occurs always at the area of Necking. The inclined curve with decreasing stress indicates the beginning of necking.

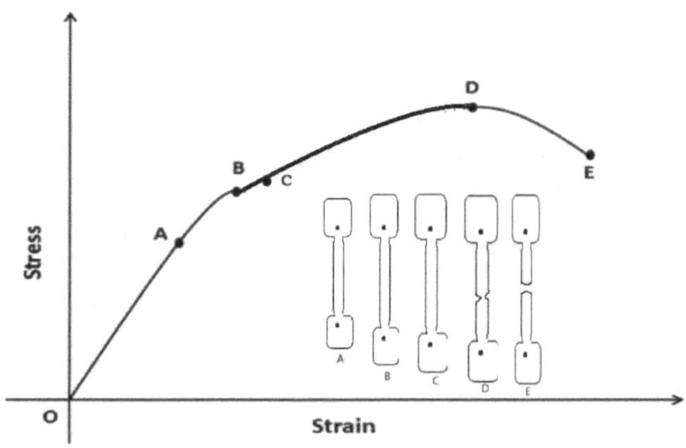

Fig. 4.19

(Necking a failure of a specimen in tension test.)

4.10 Plastic Deformation-

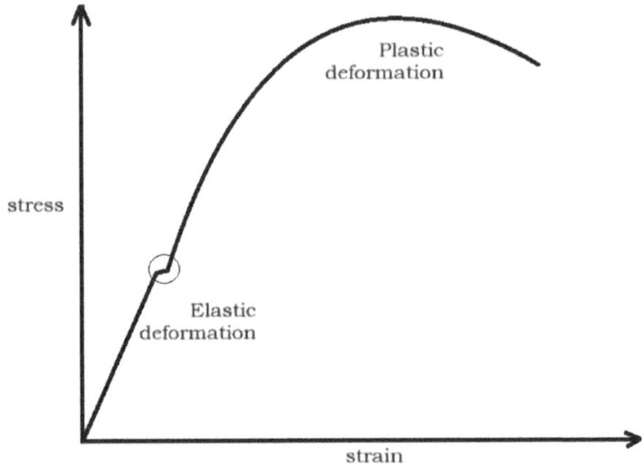

Fig. 4.20

In uniaxial tensile, initial part is a elastic deformation followed by plastic deformation. Stress strain curve of ductile material shows that up to a yield strength, the deformation was elastic, but, then plastic deformation occurred which is permanent deformation. At any point of plastic deformation, if material is unloaded, it shows permanent plastic strain present in the material even at zero stress strain remain in the material, so that the part of curvature transition occur from elastic to plastic was studied. Behavior of internal structure due to plastic deformation studied by X ray diffraction. X ray diffraction studies show that the crystalline order in the solid is not lost during plastic deformation, even though more imperfection are introduced. The atom moments are such that the crystal structure remains the same before and after plastic deformation and undeform. Structure that is crystalline order does not change after deformation due to plastic deformation, external surface is changed without any change in crystalline order.

The change in the external surface was studied by various model by the different people. Two basic models were studied successfully mainly involved for the change in external surface. The model of plastic deformation called slip and twinning models.

Slip is a shear deformation that moves atoms by many inter atomic distances relative to their initial positions as shown in figure. Steps are created at the surface of the crystal during slip but the orientation of all part of the crystal remains the same before and after the slip.
Twinning on the other hand changes the orientation of the twinned parts as shown in figure. The movement of an atom related to its neighbors is only a fraction of an inter atomic distance. The slip model of deformation is the common mode in the many crystals at ambient and elevated temperatures.
The term twinning is the plastic deformation that takes place along two planes due to a set of forces applied to a given material. It has been observed that a metal usually deforms by twinning only if it is unable to slip. Deformation produced by twinning is small. It has been observed that it may be produced either by mechanical deformation or annealing mechanical deformation is due to shock loading whereas annealing is due to cold work metal.

Mechanism of external shape change- Careful examination of the surface of a deformed crystal under the microscope shows groups of parallel lines that correspond to steps on the crystal that have sheared concerning one another is surface steps.
It is generally found that the slip planes are the plane of greatest atomic density and the slip direction is the closed-packed direction within the slip plane. Since the plains of the greatest atomic density are also the most widely spaced planes, in the crystal structure, therefore the resistance to slip is generally less for these planes than for any other set of planes. The slip plane together with the slip direction is called the slip system. FCC presents 12 slip systems including metals like copper, aluminum, nickel, and silver but slip in BCC that is 12 slip system occurs along the plane of shortage shortest burger vector as well there are no close-packed planes in the BCC crystal structure. A slip system in BCC requires heat to activate. Some BCC materials for example α-Fe can contain up to 48 slip systems, and HCP structure exhibits 3 distinct slip systems. The value of sheer stress required for such a moment of atom in a perfect lattice called theoretical shear strength is approximately equal to the value of shear modulus divided by 6G/6, however in actual practice, the value of shear stress required to produce a slip is found to be at least 100 times smaller than the theoretical value. This means that the mechanism responsible for slip is not due to the body sharing of planes of atoms but due to some other cause.
Another study that indicated external shape change was the microstructure of deformed metals. (independent study).

So, x-ray diffraction observation and microstructure observation both indicated the slip model of deformation which leads to a change in the shape of the material, a change in the shape of the crystal without the change in the internal crystal structure.

4.11 Creep-

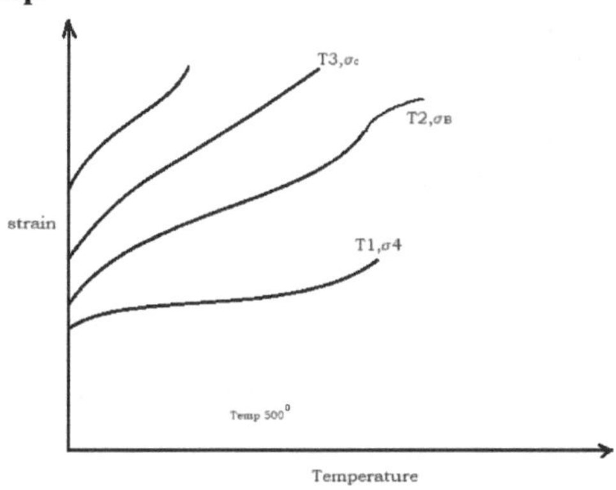

Fig. 4.21

Creep is a time-dependent continued plastic deformation at constant load or stress. The continuous deformation of a metal under a steady load is known as creep. The study of creep is important for metals used in high-temperature applications, such as heat exchangers, furnace linings, boiler baffles, turbines, etc. A few exceptions like lead, glass which creeps can also at constant room temperature. As a result, the

material undergoes a time-dependent increase in length, which could be dangerous ----
------------------.

Creep is a high-temperature phenomenon.

	Tm0C	Tm K	T roomK/ T melting K
Al Spring 1	660	933	300/933=0.32
Pb-Sn Solder Spring 2	183	456	300/456=0.66

Spring 1 Remains as it is whereas spring 2 elongate after time interval indicates elongation is a function of time.
For constant temperature, creep strain increases with increasing stress.
For constant stress, creep strain increases with increasing temperature.
So at room temperature High creep resistance is observed with the metal having a high melting point. The test for pb- sn solder is high temperature test but not so high temperature test for aluminum.Creep is a significant at temperature greater than about 0.5 Tm.
T greater than 0.5 Tm.
A coarse grain metal has higher creep resistance than a fine grained metal.
Creep test is a very essential test to predict the working life of some members or machine components which are subjected to creep. It is present in metals like iron nickel, copper, and their alloys at increased temperatures. Metals like tin, lead and their alloy also creep at room temperatures. It has been observed that some organic metals such as plastics and rubber are very sensitive to creep.
Creep in metals is caused by slip occurring along crystallographic directions in the individual crystals together with some deformations of grain bonding most of the deformation in the case of metal is non-recoverable. Generally performed by applying a static load to one end of a specimen at a constant temperature. The axial deformation is read periodically, throughout the test, a curve is plotted between extension that is strain along the vertical axis and time along the horizontal axis. The creep curve representing the extension or strain around the vertical axis and the time on the horizontal axis under application of load. The creep curve indicates the following three different stages of Creep-
creep is simply a tensile test performed at high temperature with the help of equipment first electric furnace with suitable temperature control
2nd extensometer
3rd loading device.

The load is applied to a specimen by using direct weights. Total creep or percentage elongation is plotted against time. This curve is called a creep curve.
A typical Creep Curve-
Keeping constant the stress, and strain is a function of time.
1] The creep rate or strain rate was initially quite high and gradually becoming less.

2] Creep rate is constant known as steady rate.
3] The creep rate again rises and material breaks.
Constant strain rate is important for the creep life of the material. Iss= de/dt- steady rate or creep rate.
Creep rate is considered to be high high-temperature phenomenon that is the melting point of material.
Primary stage- In this part curve indicates rapid extension of the specimen
secondary stage- the most important stage, creep is constant rate which has maximum practical utility.
The tertiary stage indicates rapid extension of sesame which finally leads to the failure of the material.
Stages One and three are for a short time as compared to stage 2, which represents the entire life period of the specimen which is of great importance because the lifetime of a machine component depends on the rate of this extension.
The value of creep increases with the increase in temperature as well as stress. Therefore, the material to be used at high temperatures must have a high melting point.

Role of Materials Selection in Design-

Material selection is a step in the process of designing a physical object. In the product design, the main goal of material selection is to minimize cost while meeting the product performance goals, such as mechanical, thermal, physical, and electrical properties. The task of selecting the right material for a given engineering design can appear to be very great in amount. The job of sorting through the options to arrive at the optimum selection requires a systematic approach based on an understanding of the nature of material science and engineering. Material selection is the final practical decision in the engineering design process and can determine the design's ultimate success or failure. The systematic approach to material selection is the use of the five material categories such as metals, ceramics and glasses, polymers, composites, and semiconductors.
Metals, ceramics and glasses, and polymers are associated with atomic bonding, composites involve a combination of two or more materials among metals, ceramics, and glasses, polymer. The first four categories are structural materials. Semiconductor is electronic material.
Material properties are used as the design parameter by the engineer to specify quantitatively the material requirement of that design.
Processing of materials can have a strong effect on their design parameters. Examples of selection within each category of structural materials include a discussion of the competition among the four different types.
Then the selection of electronic, optical and magnetic materials, includes semiconductors. Finally, the effect of engineering materials on our environment.
The sequence of choice necessary to make a final selection of metal as the appropriate type of material for a commercial gas cylinder [it's a container that must be capable of storing gases at pressure as high as 14 MPa for an indefinite period.] semiconductor cannot be considered, Polymers are also rejected because of their low strength. Of the

three common structural materials (metals, ceramics and polymers), ceramics can withstand service load by fail ductility.

The use of such brittle material in pressure-containing designs can be dangerous.

Metals provide sufficient strength and ductility to serve as excellent composites.

The next criterion is fabrication cost. Reduced weight justifies the cost.

Even considering commercially available, moderately priced alloys with acceptable mechanical properties can be listed for selection.

In making the final alloy selection, property comparisons must be made at each step in the path.

Superior mechanical properties have power and influence over the selection at certain junctions but offer the most cost. Mechanical performance typically focuses on a tradeoff or balance between the strength of the material and its deformability.

CHAPTER 5

Thermal Expansion-

5.1 Thermal expansion is the tendency of matter to change its shape, area, volume, and density in response to a change in temperature.
Temperature is a function of the average molecular kinetic energy of a substance. Molecules start to vibrate and move more and usually maintain a greater average separation.
In simple words, thermal expansion is the general increase in the volume of a material as its temperature is increased. Most matter expands when heated and contracts when cooled- this is the principle of thermal expansion.
There are three types of thermal expansions depending on the dimensions that undergo change and that are linear expansion, areal expansion and volumetric expansion.
α α 1/ bond strength and α α 1/ melting point
Linear expansion is the expansion in the length of the substance which is subjected to increasing temperature. Linear expansion mostly takes place in solids.
$\Delta l/ L = \alpha \Delta T$; where α is the coefficient of linear expansion, $\Delta l/ L$ is the fraction of length change, ΔT is the temperature change.
The ratio of expansion is directly proportional to ΔT.

Materials		α [mm/mm0C] x 106 at 27 0 C
Metals	Aluminium	23.2 x 10 -6 mm/mm0C [6600C]
	Copper	16.8 x 10 -6 mm/mm0C [6600C]
Ceramics	Al2O3	8.8 x 10 -6 mm/mm0C [20720C]
Polymer	Nylon	30 x 10 -6 mm/mm0C [268.80C]

α ceramics< α metal< α polymer
The thermal expansion coefficient of ceramics and glasses are generally smaller than those for metals, which in turn are smaller than those for polymers.
The ceramics and glasses have higher bonding energies associated with their ionic and covalent type bonding resulting in more symmetrical energy well with relatively less increase in inter-atomic separation with an increase in temperature.
For areal linear expansion, $\Delta A/A= \beta\Delta T$; where ΔA/A is the fraction of area change with change in temperature ΔT, β is the coefficient of areal expansion.
For volumetric linear expansion, $\Delta V/V= \gamma\Delta T$; where ΔV/V is the fraction of volume change with change in temperature ΔT, γ is the coefficient of volumetric expansion.
The coefficients of linear, areal, and volume expansion are related to each other.
It is found that, $\gamma= 2\beta=3\alpha$.

An increase in the temperature increases the heat of the material which leads to the increased thermal vibration of the atoms in the material. This increases the average separation distance of adjacent atoms increasing its vibrational energy.

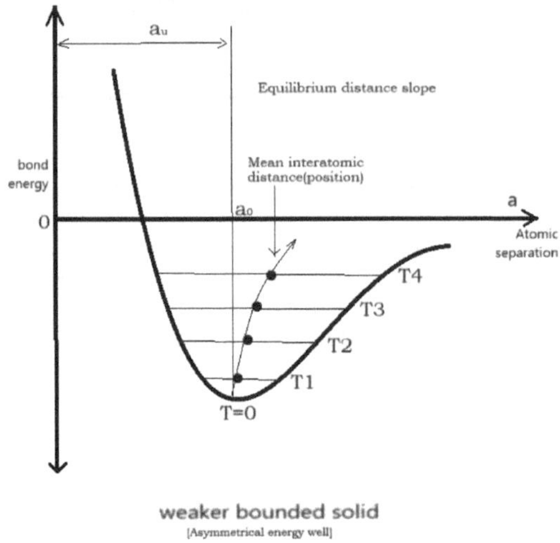

Fig. 5.1

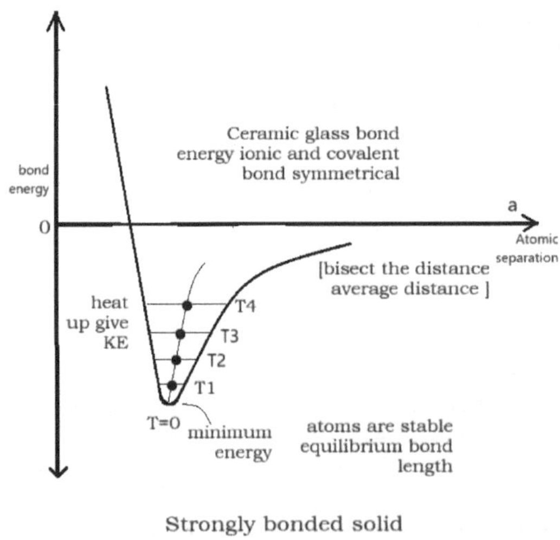

Fig . 5.2

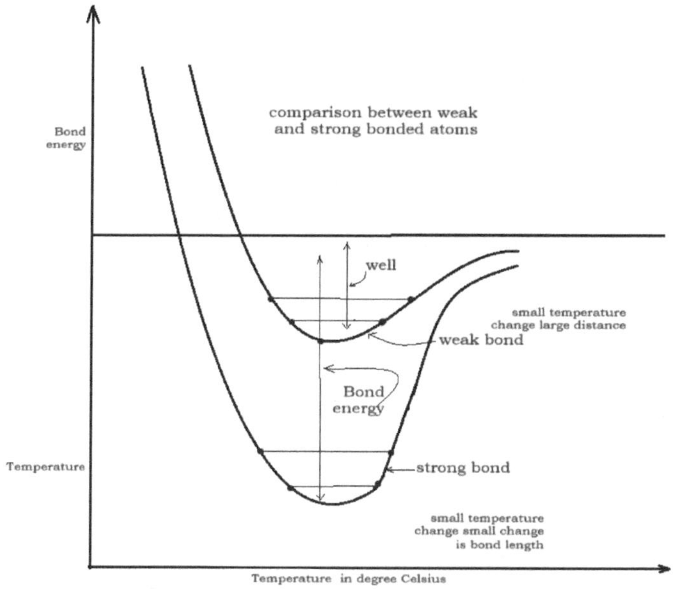

Fig. 5.3

Railway track buckled as it heated up perhaps seasonally expands and bucking behavior is observed. To avoid this, real material interacts with temperature change. So, in the bridge, thermal expansion joints are designed exactly so that bridge does not buckle and can expand at its highest temperature without failure.
Material behavior with change in temperature with microscopic deformation describe the coefficient of linear thermal expansion.

The overall dimension of the material in a given direction will increase with increasing temperature T.

Thermal expansion coefficients of ceramics and glasses are generally smaller than those for metals, which are in turn smaller than those for polymers. The difference due to energies. The ceramics and glasses have higher bonding energies associated with their ionic and covalent type bonding results more symmetrical energy well with relatively less increasing interatomic separation with increased temperature.

The coefficient of thermal expansion is inversely proportional to bond strength of the material and hence melting point.

Problem- 0.1 m long Al2O3 furnace tube is heated from room temperature 25 deg C to 10000 deg C. Calculate the increase in length.

5.2 Elastic Modulus- (Young Modulus)

It is the ratio of stress, below the proportional limit to the corresponding strain. It is the measure of rigidity or stiffness of a material. The greater the modulus, the stiffer the material or the smaller the elastic strain that results from the application of a given stress.

In the stress- strain curve, region within the elastic limit is important ro==for structural and manufacturing sectors, since it describes the maximum stress a particular material can take before being permanently deformed.

Modulus of elasticity is simply the ratio between stress and strain. Elastic modulus is of three types-

Young modulus= normal stress/ longitudinal stress

Shear modulus= Shear stress/ Shear stress

Bulk modulus= normal stress/ volumetric stress

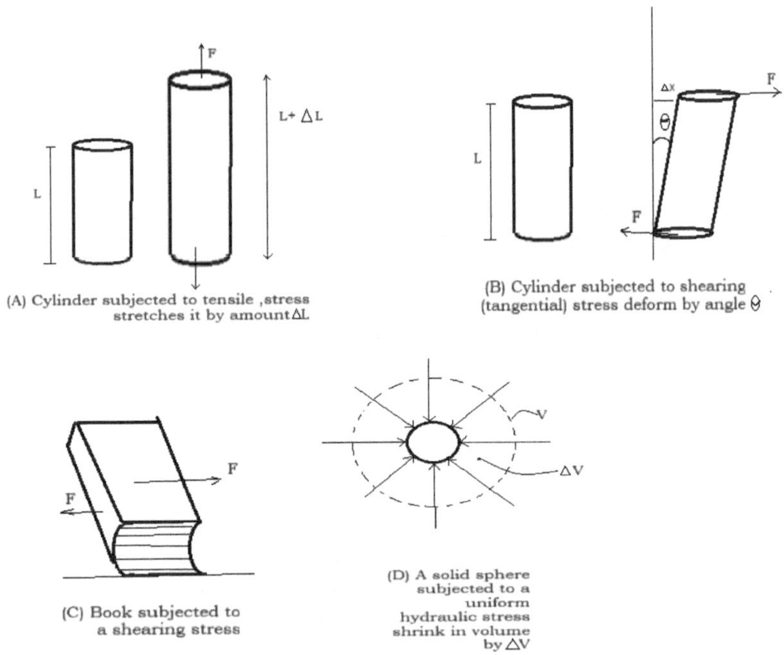

Fig. 5.4

5.3 Hook's law –

It states that for small deformities the stress and strain are proportional to each other stress is proportional to strain.

stress = Kx strain;

where K is constant of proportionality which is called as modulus of Elasticity.

NASA national aeronautics and Space Administration Space Transportation Systems where demanding thermal insulation requirement such as provide outer surface, resist surface thermomechanical load, resist moisture, protect against temperature. Hardening is an important factor in shaping metal by cold working.

The elastic modulus is directly related to the derivative of the bond energy curve near the bottom of the wheel and it follows that dip for the energy well, the larger the value of that derivative and hence the greater will be the elastic modulus. The stronger bonding associated with deeper energy well corresponds to higher melting point.

Correlation of bonding strength with material properties-

Weekly bonded solids show low melting point low elastic modulus and high thermal expansion coefficient. Stronger bonded solids show high melting point, high elastic modulus low thermal expansion coefficient.

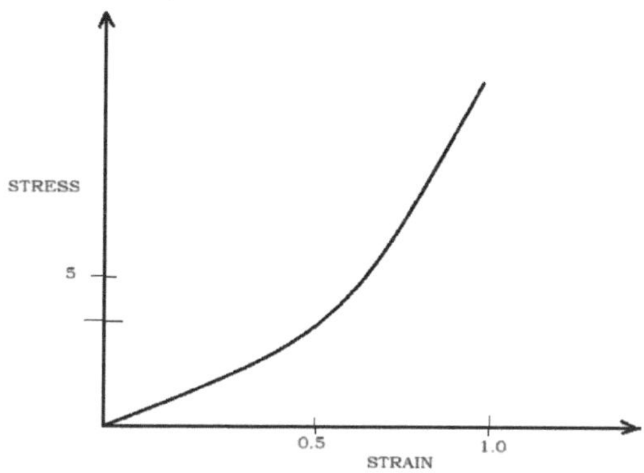

Fig . 5.5

Young Modulus majors the resistance to the separation of adjacent atom that is inter atomic bonding forces to be deformed elastically when stress is applied to it.

5.4 Melting point of metals-

One of the most important characteristics of a metal is its melting point and also known as melting temperature. Melting Point or temperature at which the solid and liquid form of a pure substance can exist in equilibrium. Adding more heat to it will cause it to further liquefy. cooling it will cause solidification. Melting Point of metal is important when metal is processed or used for metal components at higher temperature.

When more energy is needed, it means higher the melting point or boiling point. Metals are of large lattice structures electrostatic forces of the metal are large which require large amount of energy to break the bonds and so melting point of metal is high. Bond

strength leads to high melting point. The most common metals used in manufacturing are chosen based on their melting points in addition to other factors such as strength and corrosion resistance.
Melting point of various metals-
aluminum 660 degree Celsius
lead 327 degrees Celsius
brass 990 degree Celsius
copper 1084 degrees Celsius
wrought iron 1593 degrees Celsius
cast iron 1204 degrees Celsius
stainless steel 1510 degrees Celsius
steel 1540 degrees Celsius
tungsten 3400 degree Celsius.

Melting point of a solid indicates the temperature to which the material must be subjected to provide sufficient thermal energy to break its coefficient bonds. The secondary bonding is a weak link that causes the material to lose structural rigidity approximately above 120 degree Celsius. This is not a Precise melting point, but temperature above the metal softens rapidly with increasing temperature.

5.5 Metal corrosion and its Prevention-

Metallic materials were used frequently for various engineering applications, where these metals are exposed to the environment enriched with moisture and gases or immersed in liquid like acid and alkalis, it undergoes disintegration and lead to formation of chemical compounds. This will reduce the service life of the metallic components or damage the equipment.
One of the important factors which is corrosion tolerance is an essential constraint which is to be considered before optimizing a choice of material or material selection.
Corrosion is a natural process that converts a refined metal into a more chemically stable form such as oxide, hydroxide, carbonate, or sulfide. It is the gradual destruction of materials by chemical and/ or electrochemical reaction with their environment. To produce metals from natural minerals or oars it is necessary to provide a certain amount of energy for example iron ore is a blast furnace therefore thermodynamically unstable. These metals were exposed to environment would revert to their state found in nature. Now a days, it is very severe problem and their destruction of metals due to the environment has led to form a separate study under the name Corrosion Science.
Corrosion is defined as unintentional destruction of solid body due to direct chemical or electrochemical reactions starting from the surface. Corrosion is defined as destruction or deterioration and consequent loss of a solid metallic materials starting due to an unintentional chemical or electrochemical attack by their environment.
Cause of corruption-
Except noble metals like gold silver etc, all other metals exist in nature in the combined forms such as oxides, hydroxides, carbonates, sulfides, sulfates etc. Extraction of metals from their ore requires a considerable energy to be supplied to the process. Therefore, the isolated pure metal can be considered as being in a much higher energy state than

in the corresponding or and hence they exhibit a natural tendency to return back to their lower energy state that is combined state or compound form. Thus, Corrosion can be regarded as the tendency of the metals to revert back to their more stable chemical form such as oxides, sulfides etc in which they occur in nature.

Economic importance-

Everyday Corrosion of many metallic items of common use is taking place. Corrosion is a serious problem because the life of the plant or equipment is very much reduced due to Corrosion. An engineer, whether dealing with design or process operation or may be a mechanical chemical, or civil engineer has to face this serious problem of Corrosion.

For example, chemical industries face many Corrosion problems, some of which are so serious as to Cause shutdown of plants and collapse of structures involving hazardous to human life. It is very difficult to access the exact loss caused due to corrosion. This changes from process to process and from locality to locality.

In India, the direct loss due to Corrosion per annum is estimated to about rupees 200 crores and annual expenditure in controlling Corrosion is of the order of rupees 50 crores. Thus, Corrosion is a matter of great concern due to the enormous cost involved in the replacement and maintenance of metallic parts in all kinds of applications.

The oral decision is avoiding severely corrode conditions and providing efficient protection against corrosion. It is necessary to have a basic understanding of Corrosion behavior.

Types of corrosion-

corrosion occurs in many ways depending upon the corrosive environment, these may be broadly classified into-

dry corrosion or direct chemical corrosion and wet corrosion or electrochemical or immersed corrosion.

Dry corrosion - in this type of corrosion, the metal is surrounded by gases such as oxygen, sulfur dioxide, hydrogen sulfide etc in the surrounding environment and as a result, corrosion occurs mainly through the direct chemical action of atmospheric gasses within the metal surface. Dry corrosion is a phenomenon involving direct chemical relations between metal and gases, produces two important effects on the metal such as consumption of metal and changes in the metal properties.

Three types of dry corrosion depending on the corrosive environment-

oxidation corrosion

corrosion by other gasses and

high temperature corrosion

Mechanism of dry corrosion or oxidation corrosion-

Destruction of metal due to chemical reaction occurring on the metal surface due to various atmospheric gasses. Oxidation corrosion is by direct action of oxygen on metals in the absence of moisture.

The steps are-
1.Adsorption of oxygen at metal surface
2.Association of oxygen
3.Loss of electron by metal and gain of electron by oxygen and
formation of metal oxide film on the metal surface.
4.The nature of oxide formed plays an important role in oxidation corrosion process.
5.Metal + oxygen metal oxide(corrosion product)
6.When oxidation starts, a thin layer of oxide isformed on the metal surface and nature of this layer of film decides the further action.
7.Nature of metal oxide films or types of metal oxide films-

Four types of metal oxide films
a. Porous-nonprotective
b. Non- porous protective
c. Unstable
d. Volatile

Porous/ non- protective-If the oxide film is porous and having small pores, oxygen gas can pass and the process of corrosion continues.
Eg- oxide film iron -steel
$2Mg + O_2 \quad 2MgO$

Non-Porous/ protective- If the oxide film is non-porous and do not having small pores, oxygen gas cannot pass and the process of corrosion stops.
Eg- Al, Zn, Cu, stainless steel etc.
$2Al + 3/2O_2 \quad Al_2O_3$
$2Zn + O_2 \quad 2ZnO$

Unstable/ reactive- If oxide film formed is unstable, it decomposes back into the metal and oxygen, so there is no corrosion.
Eg; gold, platinum and silver (costly metals)
$4Ag + O_2 \quad 2Ag$

Volatile oxide film-
If the metal oxide is volatile, there it evaporates. The fresh metal is exposed to air, thus, faster corrosion takes place. Eg. Molybdenum oxide (Mo_2O_3) film is volatile.
$4Mo + 3\ O_2 \quad 2Mo_2O_3$

5.6 Pilling Bedworth Rule-

This describes the porous and non -porous nature of metallic oxide which is formed during corrosion.
Pilling Bedworth ratio is the ratio of volume of metal oxide formed to the volume of corresponding metal from which oxide is created.
PBR= volume of metal oxide formed/ volume of corresponding metal.
If PBR <1; metal oxide film is porous, non- protective and corrosion takes place.

If PBR >1; metal oxide film is non- porous, protective and corrosion does not takes place.

Corrosion by other gases-

Corrosion of a metallic surface by other gasses such as carbon dioxide, chlorine etc will take place or not, it depends on their chemical affinity which is the tendency of an atom or compound to combine for the meter under consideration and the intensity of attack depends on the nature of film formed on the metallic surface.

Example AgCl film formed due to attack of chlorine on silver is protective that is non-porous which decreases the attack. When chlorine acts on tin- surface that is SnCl4 is formed which is volatile and the metal surface is exposed for further attack. This result in its gradual but complete destruction of tin metal.

High temperature Corrosion –

The scale formation on metals which takes place at elevated temperature under dry condition is known as high temperature corrosion. The rate of oxidation of metal at an elevated temperature depends on the nature of oxide layer that forms on the metal surface. As the oxide layer increases in thickness, the diffusion distance through the layer also increases. The thickness of the layer increases as the square root of time at constant temperature. X= √ DT; where X is increasing thickness, D is diffusion coefficient, T is time. Diffusion coefficient D increases with the temperature in an exponential manner. Correspondingly the oxidation rate also increases exponentially with temperature.

K = A e – E/RT

where K is rate of Corrosion, A is frequency factor, E is activation energy, R is gas constant.

Hence, as the temperature increases, problem of metal protection also becomes more serious and increases rapidly.

wet corrosion or electrochemical or immersed corrosion.

The different types of wet corrosion are

galvanic corrosion

immersed corrosion

concentration cell corrosion

pitting corrosion,

waterline corrosion,

stress corrosion,

soil corrosion etc.

Electrochemical corrosion involves formation of anodic and cathodic areas separated by distance with a conducting medium called electrolyte to maintain good electrical contact between anodic and cathodic area. A potential difference between anodic and cathodic area maintain flow of electric current anodic area is that portion of the metal surface which is corroded means oxidized or lost electrons and enters the electrolyte. Cathodic area is that portion of metal surface in which electrons from the electrolyte enters and returns to the metal. Here, metal is not affected as reduction takes place. Construction of metal by electrochemical reactions occurring on metal in presence of acquires conducting medium with the formation of anodic and cathodic areas is known as wet corrosion.

Galvanic cell corrosion	Immersed Corrosion
Electrodes are different. The electrochemical corrosion of a metal by attack of an acquires conducting medium due to galvanic cell formation is known as galvanic corrosion. Water flow through metallic pipeline and tap connected to it so slow of water to electrolyte from galvanic cell. Stressed and unstressed part on the same metal Due to stress and unstress have different behavior formed galvanic cell.	The concentration cells formed due to different oxygen Concentration on a metal and the corrosion is called differential aeration corrosion. These corrosion occur at poorly oxygenated area and other parameters are involved in this. partially imersion of metal in water (ship body partially deep in water underwater part has less oxygen and corrosion occurs at larger ratio. Partial burial of metal in ground (electricity pole is buried in ground, underground parts have less oxygen area, it will corrode faster.)

Experiments-

$4Fe + 3\ O2 \quad 2Fe2O3$

$Fe2O3 \quad Fe2O3 \quad Fe2O3\ x\ H2O$

Apple contains iron. When it is cut, iron gets exposed to atmospheric oxygen and moisture. Iron gets oxidized to develop brown coating which is called as rust.

Acid Theory-

$CO2 + H2O \quad H2CO3$- carbonic acid

M + Acid   Corrosion

Carbonic acid reacts with iron to form Iron bicarbonate

$Fe + 2CO2 + H2O + \frac{1}{2} O2 \quad Fe(HCO3)2$

Bicarbonate react with oxygen to form rust.

Comparison between dry and wet corrosion

Dry Corrosion	Wet Corrosion
This is due to direct chemical reaction.	This is due to electrochemical reaction.
It takes place in presence of atmospheric gasses.	It takes place in presence of acidic, alkylene, or electrolyte.
Rate of corrosion is slow.	Rate of corrosion is high.
Product of corrosion is oxide, chloride, sulfide etc	Product of corrosion depends on electrolyte and dissolved salts.
Anodic and cathodic areas are not formed.	There is a formation of anodic and cathodic area.

Testing and measurement of corrosion
weight loss method-
rate of corrosion R= KW/DAT
electrical resistance method- For wire
electrical resistance increases as corrosion decreases.

Corrosometer - It measures the change in resistance of a standard probe metal to corrosion product.
Methods for prevention of corrosion-
The different methods used for the prevention of metal and their alloys from the corrosive attacks of the environments can be considered under the following general headings-
Proper design and material selection- proper design should avoid the presence of crevices or narrow crack between the adjacent parts of the structure. Crevices corrosion examples are joints, couplings, nuts and bolts due to accumulation of dirt or deposit.
Selection Of materials- selection of right type of material is the main factor in controlling corrosion. Along with cost, structure, chemical properties, and environment should also be considered.

Alloying- if metals are suitably alloyed, both strength and corrosion resistance of most commercial metals can be increased. Iron can be made passive by alloying it with chromium etc.

Inhibitors- a small quantity of some substances when added to corrosive environment effectively decreases the rate of corrosion of metal. Such substances are called as inhibitors. protect anode cover by chromates etc and cathode cover by amines etc.

Protective coatings- Application of protective coatings is one of the oldest way to protect underlying material. Coated component is protected due to isolation of the component from the corroding environment.

Galvanizing- zinc coating by hot dipping means galvanizing is applied to basic metal especially iron or steel where this one is to be exposed to the atmosphere or soil.

Metal spraying- Coating On the base metal by spraying the molten metal of which coating is desired by means of a portable spraying apparatus.

Electroplating- Metal is deposited from the solution to prevent against corrode.
Paint and varnishes- the coatings of paints and varnishes are applied on metallic surface and other constructional materials for corrosion protection as well as decoration.

CHAPTER 6

Diverse material and their property

6.1 Type of materials-

The materials available for practicing engineers are-

Metals, ceramics and glasses, polymers, composites and semiconductors

1. Metals- Typical materials in public mind is steel, possess several properties and considered to be metallic.

It is strong and formed into practically desirable shapes

Its permanent deformability or ductility

Metallic luster

Good conductor of electric current

Periodic table of elements highlights various engineering alloys, including the irons and steel, aluminium alloys, magnesium alloys, titanium, nickel, zinc, and copper, brass

2. Ceramics and Glasses-
   Aluminum is a common metal, but aluminum oxide is ceramic. Al_2O_3 is stable in a wide variety of server environment whereas, metallic aluminum could have been oxidized.
   Al_2O_3 has higher melting point 2020 deg C than the metallic Al having only 660 deg C, making Al_2O_3 a popular refractory and high temperature resistance material of wide use in furnace.
   With its superior chemical and temperature resistant properties Al_2O_3 is not used for applications such as automobile engines in place of metallic aluminum.
   Brittleness- Aluminum and other metals have high ductility whereas, aluminum oxide ad other ceramics lacks in this. Therefore, ceramics are eliminated from many structural applications as they are brittle.
   Development in ceramic technology are expanding the utility of ceramics for structural applications by eliminating brittleness, increasing the strength and increasing their resistance to fracture.
   A distinction between metallic and ceramic type material is that by fairly simple process techniques many ceramics can be made in a non -crystalline form (that is atoms are stacked in irregular and random patterns). For example, glass. Most common glasses are silicates, ordinary window glass is approximately 72% silica SiO_2 by weight with

balance material of Na_2O and CaO. Glasses shows brittleness and ability to transmit visible light.

Polymers- A major impact of modern engineering technology on everyday life has been made by the class of materials known as plastics or polymers. This synthetic or human made materials are special branch of chemistry. Examples of inexpensive and functional polymer products are readily available to each of us. Polymers are long chain molecules composed of many single hydrocarbon molecules such as ethylene. Important polymer includes polyethylene, simple compound of hydrogen or carbon and oxygen like acrylics, nitrogen containing nylons, fluorine containing fluoroplastics.

Plastics have desirable ductility, lightweight, low cost alternative, lower strength, lower melting point. Out of their limitations, polymers are highly versatile and useful materials. Substantial progress has been made in past decade in development of engineering polymers with sufficient high strength and stiffness to permit substitutional for traditional structural metals.

Good example is automobile body panel and many more.

Composites- These materials involve some combination of two or more components from the fundamental material types such as metals, polymers, and ceramics. A key philosophy in selecting composite material is that they should provide best attractive properties from each component. examples is fiber glass.

Strength of small diameter glass fiber is combined with the ductility of the polymer matrix to produce a strong material capable of withstanding the normal loading required of a structural material.

Wood is an excellent example of a natural material with useful mechanical properties because of its fiber reinforced structure.

Concrete is an example of an aggregate composite

6.2 Semiconductors - polymer are highly visible engineering materials, semiconductors are relatively invisible. A relatively small group of elements and compounds has an important electrical property. Semi conduction was introduced by discussion of intrinsic elemental semiconductor such as impurity free silicon extrinsic, elemental semiconductor such as silicon doped.

Nature of added impurity determines the nature of the semiconduction either N type with predominantly negative charge carriers or P type with predominantly positive charge carriers.

The modern microprocessor is a state- of the -art application of semiconductor materials enclosed in a package of conventional structural materials.

These processes contain more than 40 million transistors on a single silicon chip diode amplifier.

6.3 Amorphous Solids-

Amorphous solids lack in three- dimensional long- range order of a crystalline material, possessed a more random arrangement of molecules, exhibit short range order over a few molecular dimensions.

Amorphous solid do not have an ordered structure or not orderly arrangement of atoms or ions in a three- dimensional structure. These solids do not have a sharp melting point and the solid to liquid transformation occurs over a range of temperatures.

Properties of amorphous Solids-

Lack of long -range order- possess small regions of orderly arrangements

No sharp melting points- glass melts over a large range of temperature.

Conversion into crystalline form- Amorphous solid, when heated and then cooled slowly, becomes crystalline at some temperature-. In ancient time, glass object were looking milky because of some crystallization taken place.

Amorphous solids are isotropic- Solids exhibit uniform properties in all directions. The thermal conductivity, electrical conductivity, coefficient of thermal expansion and refractive index have the same value in whatever direction, the properties are measured in amorphous solid.

Difference between amorphous and crystalline solids

crystalline solids	amorphous solids
The constituent particles, atoms are arranged in regular and definite three-dimensional patterns. It gives clean cleavage. They have a sharp and definite melting point these are an isotropic in their physical properties and are not identical in all direction. Physical properties such as refractive index, hardness, electrical conductivity, thermal conductivity, coefficient of thermal expansion vary.	The constraint particles are arranged in irregular three-dimensional patterns. Unclean cleavage is observed on cutting. Melting point is not definite and it melt over a range of temperature. These are isotropic means their physical properties are identical in all the direction. Example are glasses, ceramics, polymers which rapidly quenched, melts. Any given crystalline solid can be metamorphose by the rapid cooling of its melt.

Any given crystalline solid can be made amorphous by a very rapid cooling of its melt Examples of amorphous solids are glasses ceramics polymers rapidly quenched melts some examples are glass, rubber, many plastic etc. Quartz is an example of a crystalline solid which has regular order of arrangement of SiO4 tetrahedra. If quartz is melted and the melt is cooled rapidly enough to avoid crystallization, and amorphous solid called a glass is obtained. This material has the same composition SiO2, but lacks the molecular level orderliness.

Amorphous solids are isotropic amorphous solids are isotropic that is they exhibit uniform properties in all directions. Electrical conductivities, coefficient of thermal expansion and refractive index of an atmosphere of an amorphous solid have the same value in whatever the direction the properties are measured.

6.4 Biomaterials-

Biomaterials may be natural or synthetic used in medical applications to support enhance, or replace damaged tissue or a biological function. It means that biomaterials can be obtained or derived either from nature or synthesized in the laboratory using a variety of chemical processes on metallic components, polymers, ceramics or composite materials. Biomaterials are often used and or adapted for a medical application and thus comprise the whole or part of a living structure or biomedical device which performs or replaces a natural function being used for a heart wall or bio active with more interactive functionality. Biomaterials include metals, ceramics, glass, and polymers found in things such as contact lenses, pacemakers, heart walls, orthopedic devices and much more. They also used in every day in dental applications, surgery, and drugs.
Biomaterials sometimes act like temporary structure to replace or repair and maintain the structure of organs as it tends to exhibit certain limitations such as material associated infections, mechanical failure of materials and reactions due to implanted materials.

<u>Bioactivity</u>- The ability of a biomaterial to induce a physiological response that is body's automatic reaction to a stimulus such as heart rate, respiration that is supportive function and performance is known as bioactivity. The term refers to the ability of implanted materials to bond well with the surrounding tissues related to bones.

<u>Self- assembly</u>- it describes the spontaneous arrangement of particles or atoms or molecules without the influence of any external forces. Large group of such particles are known to assemble themselves into thermodynamically stable structurally well-defined arrays, quite similar to one of the 7th crystal systems found in metallurgy. For example FCC, BCC. Molecular self- assembly is found widely in biological systems and provides the basis of a wide variety of complex biological structures. Thus, self-assembly is also emerging as a new strategy in chemical synthesis and nanotechnology.

<u>Biocompatibility</u>- Biocompatibility is related to the behavior of biomaterials in various environments under various chemical and physical conditions. The term refers to specific properties of a material without specifying where or how the material is to be used. The term refers as how biomaterials interact with the human body and eventually how those interactions determine the clinical success of a medical device (Such as pacemaker or hip replacements).
Biocompatibility depends upon the
composition of the implanted material,

surface of the implant
mechanism of fatigue and
there are several other reactions possible.

Some of the most commonly used biocompatible or biomaterials are polymer due to their inherent flexibility and adjust or correct mechanical properties. Medical devices made of plastics are often polycarbonate, polyvinyl chloride, polyethylene and even polypropylene.

To ensure biocompatibility, there are a series of Regulated test that materials must pass to be certified for use. These include
biological reactivity test and
biological evaluation of medical devices.

The main objective of biocompatibility test is to quantify the acute and chronic toxicity of materials and determine any adverse effects during working conditions.

Thus, test required for a given material are dependent on its end use (that is blood, central nervous system etc.).

Biomaterials must be compatible with the body and there are often issues of biocompatibility, which must be resolved before a product can be placed on the market and used in a clinical setting.

Mechanical properties in addition to a material being certified as biocompatible, biomaterials must be specific to their target application within a medical device. This is especially important in terms of mechanical properties which gives the biomaterial behavior.

Young modulus- one of the most relevant material parameter is the young modulus E, which describes materials elastic response to stress. Young's modulus of tissues and the device that is being coupled to it must closely match for optimal compatibility between device and body whether the device is implanted or mounted externally.

Matching the elastic modulus makes it possible to restricted movement at bio interface between implant and tissue as well as avoiding stress concentration that can be laid to mechanical failure.

Tensile and comprehensive strength which quantify the maximum stress of a material can withstand before breaking and may be used to set stress limits that a device may be subject to within or external to the body. Depending on the application, it may be desirable for a biomaterial to have high strength so that it is resistant to failure when subjected to load, however in some other applications, it may be beneficial for the materials to build low strength.

One application where high strength is not desirable is in neural probes- if high strength material is used tissue will fail as Young's modulus of tissue is of the order of 500 pa.

Temperature fluctuation experienced by implant biomaterial -example is in dental implantation, ductility is important. Must be ductile for a similar reason that the tensile strength cannot be too high, ductile allows the material to bend without fracture and also prevent the concentration of stresses in the tissue when the temperature changes.

Self- Assembly. Self -assembly is the most common term used in modern scientific community to describe the spontaneous aggregation of particles without the influence of any external forces. Molecular Self Assembly is found widely in biological systems and provides the basis of a wide variety of complex biological structures this includes an emerging class of mechanically superior biomaterials based on microstructural features. Thus, self- assembly is also emerging as a new strategy in chemical synthesis and nanotechnology.

6.5 Biomaterial Applications-

Biomaterial is essentially a material that is used and adopted for a medical application. Material can have a function such as being used for heart wall, hip joints, dental applications, surgery and drug delivery. Biomaterial are used in joint replacement, bone plates, bone cement, dental implants for tooth fixation, blood vessel, heart walls, skin repair devices. Biometrics must be compatible with the body and there are often issues of biocompatibility which must be resolved before a product can be placed on the market and used in a clinical setting.

Intraocular lenses- By the age around 75 more than 50% of population suffer from cataracts. Good vision is generally restored almost immediately after lens is inserted, made of elastomer and other materials.

Heart valves- It is fabricated from metals, elastomers, fabrics. It must Not react with chemicals in body. It is attached by a polymer mesh more than 45000 replacement walls are implemented every year in the US.

Dental implants- small titanium fixture that serves as the replacement for the root portion of a missing natural tooth. Implant is placed in the bone of the upper and lower jaw and allowed to bond with bones. It is Bioinert that is no reaction occurs with tissues.

Hip replacement- Common medical practice using biomaterial is corrosion resistance high strength metal alloys. Very high Molecular weight polymers are used as a thermoset plastic.

Polymers are used across almost every sector including packaging, medicals, textiles, consumer product transportation, electronics etc. Polymers are unique materials with properties such as lightweight, toughness, flexibility, transparency and they are cheaper too; but their mechanical properties are not adequate for many high- end applications such as load bearing applications. For example, for making airplane, automotive parts. For those applications, polymers alone cannot be used. If we Compare polymers with metal, polymer possess low stiffness and strength. Their modulus is 100 times lower and the strength is five times lower than the metals, which limits their application as high strength materials. It is found that if polymer matrix is fully filled with fibrous materials, the resulting material possess very high strength and excellent mechanical properties. It can be used for load bearing and other high- end applications. Such materials are known as composites. A composite is a material composed of two or more distinct constituents or phases separated by an in identifiable interface.

The idea of composite is not new. In nature, composites exist for several years.

Wood is a composite. It is made of straw and mud.

Our teeth, skin, bones, stones, concrete, all are exapmles of natural composites.

Most composites have two constitutents materials and have different chemical and physical properties.

Matrix- surround and hold reinforcement fillers.

Fillers- . Strengthens and Stephens the Matrix material.

The two materials work together to give the composite superior properties. Matrix surrounds and binds together the filler material, which provides strength and stiffness to the matrix material.

Polymer composite classification-

Polymer composite are classified in three categories based on the type of reinforcement that is added in the matrix.

Particulate composite- In particular composite, size of the reinforcing filler is a small and their shape is spherical. Aspect ratio of three type of filler is close to one.l/d=1.

Discontinuous fiber composite- In this case, short fibers are used as reinforcing fillers. Aspect ratio of short fiber is in the range from 1-10; l/d >1-10.

Continuous fiber composite- in this case the length of the reinforcing fiber is almost equal to the length of the product or more l/d >>>>1-1000. These continuous fibers can be oriented in different directions. Depending upon the direction of load bearing requirement in the composite, orientation of filler is decided.

Continuous fiber composites are much stronger than the other two types of composite because load is directly hold out by the fiber whereas in other two, load is hold by the matrix and then the filler.

Main function of filler in case of particular and discontinuous fiber composite is to improve modulus of composite. Strength of the composite will mainly depend on the chemistry between the filler and the matrix. If interfacial attraction between these two is good, composite will show good strength. Is not good then the modules will be improved but not the strength.

Fillers- Fellers are those which fall under particulate, discontinuous and continuous category. Spherical and irregular shaped particles such as carbon black, fume silica etc are particulate type of fillers and these fillers are widely used in industry especially carbon black and silica.

Glass, carbon fibers are used to make continuous composites. These composites have very high strength and can replace metals.

When continues fibres are chopped and added in the polymer matrix, the resulting composite is called discontinuous fiber composite.

Other than these, cellulose fiber, coconut shell powder, jute, silk fiber are also added in the polymer matrix as reinforcement.

Nanofillers in the polymer matrix-

A recent trend is to add nanofillers in the polymer matrix and such type of composites are called as nanocomposites.

polymer + nano filler = polymer nanocomposite

Examples of nanofiller are cellulose nanocrystal or nanofibers, carbon nanotube or nano fibers, carbon black etc. These fillers are called nanofillers because dimensions is in nanometer (nm). For example, carbon nanotubes diameter is 5- 20 nm. Because Of their nano size, surface area of these filler is extremely high which is very important to achieve could reinforcing properties.

Polymers vary in their chemical structure and offer different properties, therefore selection of polymer matrix made based on application and property requirement.

Nano composite can be polymer based composite due to small size of the nanoparticle of the structural unit and high surface area to volume ratio. Materials have large surface area but a very small volume it means S is large and V is small., surface phenomena get into action to describe the process when we talk is no more important and density is also more. Due to small size of this radical units and surface to volume ratio these narrow composites are found in which nanoparticles include in nanocomposites are finding large utilizations.

Various shapes are-

spherical particle=3/r

fiber particle=2/l+2/r

layered matrix=2/t +4/l;

where r,l,t represent the radius length and thickness represent respectively. Surface energy increases as we move from macro to nano.

In macro surface area, Surface area same but volume increases. So gravitational force starts playing important role.

Nano composite addition by 5% nanofiller into the matrix itself brings a drastic improvement in the thermal and mechanical properties. Filler can be-

zero dimensional that means nanoparticle or

one dimensional that means nano wire or nanotube or

two dimensional that means thin film coating or

3 dimensional that means embedded network or copolymer.

These are some of the fillers which are possible by nano materials which are getting integrated into the composite.

<u>Matrix-</u>

Thermoplastic polymers- Thermoplastic can withstand more heat than others. It will Melt or soften when heated and can be moulded in any shape or remoulded, holding their shape when cooled.

Example polyethylene, polypropylene etc

Thermoplastics can be used to make composites. Most consumer products are made of thermoplastic composites.

<u>Thermoset polymers</u>- Thermoset have cross linked and do not soften on heating and it cannot be reshaped. Thermosets are highly cross linked materials so they do not soften at high temperature. Thermoset composites remain stable in high and dynamic temperature. Therefore, suitable for applications where extremely high modules and high temperature stability is required, for example epoxies, polyesters, phenolics etc.

<u>Elastomers</u>- Elastomers are polymers with a limited amount of cross linking, which allows them to stretch bend and then snap back to their original shape. For example natural rubber, synthetic polybutadiene etc.

Elastomers are flexible low strength material. Their strength can be improved by incorporating fillers like carbon black, silica etc. So elastomers can be used as metrics material to develop polymer composite with flexibility and high strength.

Synthesis of polymer composites- Incorporation of filler in the polymer matrix to develop composite so these methods are -melt mixing, solution blending, in situ copolymerization. All three methods are used to make particulate and discontinuous fiber composites. Continuous fiber composites have direct product manufacturing process and there is no synthesis method solution blending melt mixing.

Melt mixing is the simplest method to mix filler in the polymer filler and polymer are taken together in an extruder and high shear is applied by the screw of extruder to disperse and distribute the filling particles.

This technique is particularly important for the preparation of insoluble and thermally unstable polymers, which cannot be processed by solution or melt processing. In insitu polymerization, monomers are polymerized in presence of fillers. Depending on the required molecular weight and free radical, ionic and ring opening polymerization.
To achieve composite material with designed properties, these two factors must be taken care of.
One dispersion and distribution of filler- dispersion and distribution of filler particles in the polymer matrix where filler in polymer matrix is always present in the form of aggregates. These aggregates are also called as agglomerates.

Agglomeration is a tendency of high surface area fillers where integral filler particles join and form a big lump. Due to high surface area of filler particles, surface contact area is very high therefore particle interaction is more and because of high surface area, filler particle like to form aggregates. So if the filler is present in aggregate form, the composite will show poor performance therefore it is important to open these bundles into individual particles. The processing of opening or dispersing fillers in individual particles is called dispersion.
If the filler particles are individually present in the polymer matrix, dispersion will be good. For better performance of the composite material both dispersion and distribution should be good. otherwise one has to compromise with the properties and performance of composite material.
Interfacial interaction between filler and the polymer matrix- this depends on two factors one surface of volume ratio or aspect ratio in case of nanobilar and
Second is chemistry between filler and the polymer matrix.
The surface to volume ratio is the amount of surface area per unit volume of an object and is inversely proportional to the size. So the materials with a high surface area to volume ration are those that are small diameter or very porous in structure. These reacts at much faster rate than the macro-materials because more surface area is available to interact with polymer.
Chemistry between filler and polymer matrix is more important for example if we try to mix hydrophobic filler in hydrophilic polymer, no matter how small the size of the filler is the resulting composite will have poor properties due to poor interfacial interaction. So to improve adhesion, fillers are usually surface treated to improve compatibility, which in turn improves interfacial interaction between the two. It also helps to reduce aggravation of high surface area filler.
To achieve polymer composite with the desired properties, dispersion or distribution of filler, their size and interfacial interaction between filler and matrix must be good. By choosing an appropriate combination of reinforcing filler and matrix material and processing method, we can produce polymer composite with a desired properties that exactly fit the requirement for a particular application. That is the reason why demand of polymer composite with superior properties is going and hence it is a constant quest of finding more advanced materials.

6.6 Polymer nano composite materials-

Composites- Composite are materials made from two or more constituent materials with significantly different physical or chemical properties that when combined produce a material with characteristic different from the individual components.

Nanocomposite- nanocomposites have nano size in gradients.

Importance of nanocomposites- Superior overall properties compared to constituent properties.

Nano composites have high surface to volume ratio of the reinforcing phase and high aspect ratio. Aspect ratio= width / height compared to conventional composite materials.

Atoms have averaged radius of about 0.1 nanometer nanospherical particle has diameter of 100 nanometer S / V = 100 / 1 which is equal to efficiency of nano substance.

Large ratio opened many possibilities for creating new materials and facilitating chemical processes. In conventional, materials most of the items are not a surface, they form the bulk of the material such as powder or plate and sheets.

The rate of fiber is to take the maximum load and the role of Matrix is to hold the fiber together and transmit the load of the fibers. Between fiber and the matrix is very critical.

Location of nanocomposites- Lithium ion batteries, fuel cells, aerospace marine applications and many more.

Main properties of nanocomposites- mechanical properties, increased ductility, mechanical properties strength, optical properties like light transmission characteristics.

Polymer nanocomposites- One of the component is polymer to which added nanometer.

Different types of nanocomposites- ceramic matrix nanocomposites

metal matrix nano composites and polymer matrix nano composites.

Applications- automobile, aerospace injection molded products, coatings, adhesive, packing materials, microelectronic packaging, optical integrated circuit, drug delivery, sensors, membranes, medical devices consumer goods etc

Typical fibers and their tensile properties-

Important fiber materials are glass, carbon CNT long or short fiber carbon nano fibers, boron silica and synthetic fibers such as aramid Value of E is highest for carbon and boron. Thermoplastics- polypropylene polyester, polycarbonate etc

choice of the matrix depends on the application service condition and cost

comparison of FRP fiber reinforced plastics with high strength steel and aluminum alloys for aircraft application

if matrix polymer release volatile material or gasses during curing process that will introduce defect mostly polyester and epoxy resin is selected. Some wounded at home temperature and pressure compared to other thermoset.

Tensil properties of epoxy based composites and cell strength is improved by adding fiber. Thermosets provide high strength, excellent adhesive to the fibers, thermal resistance, chemical resistance.

6.7 Polymer nano composites-

Introduction- Nano technology structure hundred nanometer or smaller but formed on a nanometer scale with widespread applications. Lot of research is going on in finding nanometer and nanocomposites. Nanocomposites are material made by mixing of two different materials with different properties but one of them having precise in the range of Nano scale. Polymers nanocomposites which are made up of polymer metrics and nanometer. By inserting the nanometric inorganic compounds the properties of polymers improved and hence this has a lot of applications depending upon the inorganic material present in the polymers. They have high thermal, electrical and mechanical properties.

Nano composite- A nano composite is a multi phase solid material where one of the phase has one or 2 or 3 dimensions of less than 100 nanometer or structure having nanoscale repeated distances between the differences that make up the material. It includes porous media, Colloids, gels, and Copolymers.
The solid combination of a bulk matrix and a nano dimensional phases differing in properties due to dissimilarities in structure and chemistry.
There are three major classifications of Nano proposites and table shows different examples in each category.
Ceramic matric nanoComposite
metal matrix nanocomposite
polymer matrix nanocomposite
Different types of nanocomposition-
The mechanical, electrical, thermal, optical, electro chemical, catalytic properties of the nanocomposites will differ distinctly from that of the component materials.

Polymer Nano composite - nano materials getting integrated into the polymer matrix for producing composites.
All the polymers can be used as a structural materials without reinforcement, their uses get limited due to insufficient mechanical properties. Relatively low strength coupled with low impact strength is a characterization of polymer materials.
Polymers nano composites represent a new alternative to conventionally filled polymers. polymer nano composites are materials in which nanoscopic inorganic particles, typically 10- 100 A0 unit in at least one dimension, are dispersed in an organic polymer matrix such as polyacetylene or polyethylene ethylene, polythiophene.
Because of the nanometer size of the inorganic filler, nanocomposites exhibit mark improvement in the properties when compared to the pure polymer or their traditional composites.
Enhancement in the properties- Polymer nano composites has revealed clearly the property advantages that nanomaterial additives can provide in comparison to both their conventional filler counterparts and base polymer. Properties which have been shown to undergo substantial improvements include
mechanical example given strength, modulus and dimensional stability
improved solvent and heat resistance and decreased flammability.
Decreased permeability to gases, water and hydrocarbon.
Thermal Stability and heat distortion temperature
chemical resistance,
surface appearance,
electrical conductivity,

optical clarity in comparison to conventionally filled polymers

Applications-

nano technology is deeply embedded in the design of Advanced Devices for electronic and optoelectronic applications. The dimensional skill for electronic devices has now entered the nano range. The utility of polymer based nanocomposites in these areas is quite diverse involving many potential applications and have been proposed for their use in various applications.

They are used in chemical sensors, electro catalysis, batteries, smart windows and memory devices.

Another potential application includeshotovoltaic cells and photoiodiodes, super capacitors, printable conductors, light emitting diodes.

The electrical conductivity of carbon nanotubes in insulating polymers has also been a topic of considerable interest. Th potential Applications include electromagnetic interference shielding, transparent conductive coatings, electromechanical actuators. Conjugated polymers with various nanoscale fillers inclusions have been investigated for sensor applications including gas sensors, biosensors, and chemical sensors. The nanofillers employed include metal oxide nanowires, carbon nanotubes, nano scale gold, silver, copper, platinum.

Polymer based Solar cells have the capability of being used to make cheap large flexible panels. Cadmium chalcogenides such as CDS Nanoparticles in polymers have been used to make solar cells.

6.8 Glass-

Some engineering materials lack the repetitive crystalline structure. These non crystalline or amorphous solids are imperfect in three dimensions. Glass is a non crystalline, often transparent amorphous solid that has a widespread practical, technological and decorative use in for example window planes, table ware and optics. Glass materials are excellent thermal insulator and electrical insulators and have high dielectric strength. At ordinary temperatures, glass materials are relatively strong and corrosion resistant.

Most common classes are silicates, ordinary window glass is approximately 72% silica by weight with the balance of the material sodium oxide and calcium oxide. Glass share the property of brittleness and crystalline ceramics. Glasses are important engineering materials because of the other properties such as their ability to transmit visible light as well as ultraviolet and infrared radiations. Lattice structure of non crystalline version of material is referred to as Zachariasen model in simple way it illustrates the important feature of oxide glass structure. William Zachariasen, American physicist described glass structure in early 1930 and become a standard definition for the structure of these non crystalline materials.

The building block of the crystal is retained in the glass that is short range order is retained but long range order that is crystalline is lost in the glass. Zachariasen model is the visual definition of the random network theory of glass structure which is related point lattice associated with crystal structure.

First example of non cystalline solid was the traditional oxide glass because many oxides especially the silicates are easy to form in a non crystalline state.

Silicates are the minerals containing silicone and oxygen in tetrahedral units, which are linked together in several patterns.
Rapidly cooling a liquid silicate or allowing a silicate vapor to condense on a cool substance effectively freezes in the random stacking of silicate building blocks. Many silicate glasses are made by rapidly cooling liquids, the term super cool liquid is often used with glass. Super cool liquid is the material cooled just below the melting point where it still behaves like a liquid glass which is cool to a sufficiently low temperature. So that it has become truly rigid solid.
Interatomic distance is a short range order refers to the regular and predictable arrangement of atoms over a short distance usually within 1 or 2 atom spacings.
However regularity does not persist over a long distance. Amorphous solids and glasses are materials in which atoms have SRO but no translational symmetry.
True solids are the solids which have definite melting point and definite arrangement of atoms, have definite volume. for example crystalline solids
Pseudo solids are the solids which do not melt sharply and get distorted by bending and compressing. for example amorphous SRO super cooled liquid. Silicate tetrahedron has a silicon cation at the center of tetrahedron and four oxygen anions at the four corners. Cation- cation repulsion is high. Silica Cannot form close pack structure.

As with crystalline silicates, glasses are generally moderate in cost due to the large amount of availability of elements SI and O in the element Earth's crust. For routine glass manufacturing, sio2 is readily available in local sand deposits of adequate purity. Composition of some silicate classes-

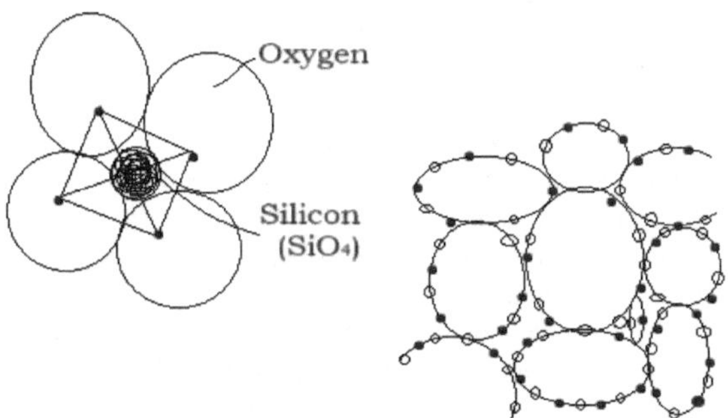

Fig 6.1
Glass materials are heated to a temperature sufficient to produce a completely fused melt. During Fabrication, the properties of glass can be greatly modified by additives or heat treatment. Examples of raw materials include cullet like crashed recycled glass, sand, soda ash, limestone and additives.
Types-
Tempered glass- Increased strength and toughered glass, safety glass. Passenger vehicle window, glass doors, tables, refrigerator trays. Processed by thermal or chemical treatment.

Borosilicate Glass (Silica Boron Oxide)- It offers superior durability, chemical and heat resistance, which make it suitable for the use in chemical processes, high powered lamps, for cookware and heat resistance product.
Soda lime glass- It is used in everyday products such as bottles, jars, drinking glasses and window glass.
Lead Glass- Increased index of refraction, relatively soft, a better electrical insulator, used for optical applications such as prism and lenses.
Vitreous Silica- Vitreous silica means glassy, with absence of significant network modifiers. It can understand service temperature in excess of 1000 deg C. high temperature crucible and furnace window are applications.

6.9 Ceramics-

Ceramics have traditionally been used in high temperature engineering applications. For the interior architecture of furnaces, silicon carbide provides good dimensional stability at temperatures up to 1650 degree Celsius, along with high resistant high resistance to thermal shock and corrosion and a low density. Ceramics are used to make pottery bricks and glass since silicon and oxygen together account for roughly 75% of the elements in the earth crust, these materials are abundant and economical. Many of the traditional ceramics fall into this category that crystalline ceramics $SiO\ 2$ based silicates. One best example in early civilization is pottery, burnt clayware that has been a commercial product since roughly 4000 BC. Pottery is a part of category of ceramic known as white wares which was commercial fired ceramics. Non silicate oxide ceramics, which includes some traditional materials such as magnesia MgO, refractory widely used in the steel industry.
WC, TiC, Al2O3 etc ceramic shows some common properties like electrical insulation, high temperature strength, high melting point, hardness etc. Application of Al2O3 ceramic in spark plug and electronic packaging etc.
Most of the ceramics cannot be used for structural purposes due to having poor tensile and impact strength. Al2O3 M pt= 2050 deg C, hardness= 2100 kg/mm3.
Bioceramics are used as dental implants and synthetic bones.
WC and TiC are used as composites bonded with a metal binder like Ni, Co, Cr, Mo etc. Such type of materials or composites bonded with a metal binder like Ni, Co, Cr, Mo etc. Such type of material or composites of ceramics and metal are termed as metal bonded carbides or cemented carbides used in cutting tools. WC has M pt= 2850 deg C, hardness= 2000 kg/mm3.

History archeologist found that human met ceramics made of animal facts and bones mixed with fine clay like materials, which was fired at temperature between 500 to 800 degree Celsius, while it is not clear what the ceramics were used for. In 9000 BC, first use of functional pottery vessels were to hold and store grain and food while firing pottery, the presence of calcium oxide containing sand combined with soda and the overheating of the pottery may have resulted in a colored on the ceramic pot. Pottery making flourished in Upper Egypt about 8000 BC. Since these ancient time the technology and application of ceramics has steadily increased, has ceramics have played major role in the progress of mankind.

Ceramic properties-

The properties of ceramic materials like all materials are dictated by the type of atoms present, type of bonding between the atoms and the way of atoms are packed. Most ceramics are made up of two or more elements or compounds for example Al2O3 is a compound metaphor of aluminum atom and oxygen atoms. Most two common chemical bonds for ceramic materials are covalent and ionic bonds due to ceramic materials wide range of properties, they are used for many applications. In general, most ceramics are hardware refractory thermal insulator, electrical insulator, non- magnetic oxidation resistant, resistance to thermal shock, chemically stable.

Ceramic Processing-

Ceramic processing is used to produced Commercial products that are very diverse in size, shape and material composition, structure and cost. Ceramics are typically produced by the application of heat upon processed clays and other natural raw material to form a rigid product.

Ceramics products that use naturally occurring rocks and minerals as a starting material must undergo special processing in order to control purity, particle size, as there play a big role in the final properties of the finished ceramics.

The next step is to form the ceramics particles into a desired shape by additional of water, additive such as binders followed by shape forming process. Some of the most common forming methods are extrusion, pressing, castings and injection moulding. After the particles are formed, there green ceramics undergo a heat treatment sintering or firing to produce a rigid, finished product.

Ceramics are classified as inorganic and non metallic materials that are essential to our daily lifestyle. Ceramics are all around us like tile, brick, plates, glass etc. Ceramics can be found in products like watches automobile spark plug phone lines airplanes electrical conductor and insulators.

Ceramic is an inorganic non metallic solid that is prepared from powdered materials and is fabricated into products through the application of heat.

CHAPTER 7

Nano composite materials

7.1 Introduction to Nanomaterials

Nanomaterials are materials that possess at least one dimension in the nano meter range (1-100 nano meter). Due to their size, these materials exhibit properties that are significantly different from their bulk counterparts, such as increased strength, lighter weight, and enhanced chemical reactivity. The term "nano" derives from the Greek word "nanos," meaning dwarf, highlighting the extremely small scale of these materials. Nanotechnology, the field that deals with these materials, has revolutionized many industries, including medicine, electronics, energy, and environmental science.

Types of Nanomaterials

Nanomaterials are broadly classified into four categories based on their dimensionality:

- Zero-dimensional (0D) nanomaterials: These materials have all dimensions confined to the nanoscale, such as nanoparticles and quantum dots.

- One-dimensional (1D) nanomaterials : These include nanowires and nanotubes, where only one dimension is not confined to the nanoscale.

- Two-dimensional (2D) nanomaterials: Graphene is a prominent example of a 2D nanomaterial, where only one dimension is outside the nanometer range, giving it a sheet-like structure.

- Three-dimensional (3D) nanomaterials : These are bulk materials that contain nanoscale structures within them, such as nanoporous materials.

7.2 Unique Properties of Nanomaterials

The unique properties of nanomaterials are largely a result of two factors: their high surface area to volume ratio and quantum effects that become prominent at the nanoscale. These properties lead to several distinct behaviors, including:

Mechanical Properties : Nanomaterials often exhibit greater strength and durability compared to their bulk equivalents. For example, nanostructured metals can be several times harder than their traditional forms.

- Optical Properties: Due to quantum confinement, nanoparticles can have unique optical properties. For instance, gold nanoparticles appear red rather than gold at the nanoscale because of their interaction with light.

- Electrical Properties: Nanomaterials like carbon nanotubes and graphene possess extraordinary electrical conductivity. They are being explored for use in next-generation electronics such as transistors and sensors.

- Chemical Properties: The increased surface area of nanomaterials often results in enhanced chemical reactivity, making them useful in catalysis, energy storage, and sensing applications.

7.3 Applications of Nanomaterials

Nanomaterials have numerous applications due to their remarkable properties:

- Medicine: Nanoparticles are used in drug delivery systems, allowing for targeted delivery of therapeutic agents, reducing side effects. They are also being explored in cancer therapy, imaging, and diagnostics.

- Electronics : Nanomaterials like graphene and carbon nanotubes are being researched for use in transistors, batteries, and supercapacitors, potentially revolutionizing the energy storage industry.

- Energy : In solar cells, nanomaterials improve efficiency by enhancing light absorption and reducing material usage. In fuel cells and batteries, they increase energy density and reduce degradation over time

7.4 Polymer nano composite materials

Composite: composite are materials made from two or more constituent material with significantly different physical and chemical properties that when combined produce a material with characteristics different from the individual component.

Nano composite: Nano composite, which have Nano size ingredients .

Importance of Nano composites: superior overall properties compared to construction properties

Nano composites have high surphase to volume ratio or of the reinforcing phase and high aspect ratio Compare to conventional composite material atoms---> average radius of about 0.1nm, nano spherical practical has diameter of 100 nm

Large ratio open many possibilities for creating new material and facilitating chemical processes. In convention material most of atoms are not a surphase, They form the bulk of material such as powder plates and sheets

Application of nano composites:- lithium iron battery is fuel cells, aerospace marine application and many more

Main properties of nano composite:- mechanical properties, increase ability, mechanical property strength, optical properties, light transmission characteristics

Polymer nano composites:- one of the component is polymer to which added Nano materials

Different type of nano composites:-

1] ceramic matrix nano composites

2] metal matrix Nanocomposites

3] polymer matrix nanocomposites

Application of polymer nano composites:- automobile aerospace, injection, molded product ,coatings adhesive, packing material microelectronics packaging ,optical integrated circuit, drug delivery, sensors, membrane ,medical devices, consumer goods etc.

7.5 Polymer Nano composites

Introduction:- nanotechnology (structure 100 nm or smaller) preferred on a nano material metre scale with widespread applications lot of research is going on in finding nano materials and Nano composites. Nanocomposites are material made of mixing of two different materials (with the different properties) but one of them having the size in the range of Nano scale polymer nano composites, which are made up of polymers matrix, are nano materials by inserting the nano matric inorganic compound. The properties of polymer improves and hence this has a lot of application depending upon the inorganic material present in the polymer. They have a high thermal, electrical and mechanical property characteristics.

7.6 Nanocomposites

Nano composites is a (solid place in form of temperature and particles) multiphase solid material where one of the phase has one, two or three dimension or less than 100 mm or structure having nano-scale repeated distance between the Different phase that make up the material. It include porous media, colloids, gel and co-Polymer. The solid combination of a bulk matrix and nano-dimensional phases differing in properties due to dissimilarities in structure and chemistry.

There are three major classification of nano composites and table shows different examples in each category

1. Ceramics matrix nanocomposites
2. Metal matrix nanocomposites
3. Polymer matrix nanocomposites

Different types of nanocomposites

Types	Examples
Ceramics	Al_2O_3/SiO_2, Al_2O_3/TiO_2, $Al_2O_3/$ SiC
Metals	Fe-Cr / Al_2O_3, CO / Cr , Fe/ MgO, Mg/ CNT
Polymer	Thermoplastic / thermoset, polymer /layered silicanes, Polymer / TiO_2 , Polymer / CNT , Polymer / Layered double hydroxide

The mechanical, electrical, thermal, optical, electrochemical, catalytic properties of the nanocomposites will differ distinctly from that of the component materials.

7.7 Polymer Nanocomposites

Nano materials getting integrated into the polymer matrix for production component although polymers can be used as structural materials without reinforcement, their usage gets limited due to insufficient mechanical properties. Relatively low strength, Coupled with low impact strength is a characterization of polymer materials.

Polymer nanocomposites represent a new alternative to conventionally filled polymers, polymer nanocomposites are materials in which nanoscopic inorganic particles, typically 10-100 A in at least one dimension, are dispersed in an organic polymer matrix such as polyacetylene (Polyethylene), polythiophene Because of their nano meter sizes of the inorganic filler nanocomposites exhibit mark improved properties when compared to the pure polymers or their traditional Composite.

7.8 Enhancement in properties

Polymer nanocomposites has revealed clearly the property advantages that nano material additives can provide in comparison to both their conventional filler counterparts and base polymer.

Properties which have been shown to undergo substantial improvements include:
1.Mechanicals eg. Strength, modules and dimensional stability.

2.Improved solvent and heat resistance and decrease flammability

3.Decreased permeability to gases, water and hydrocarbons

4.Thermal stability and heat distortion temperature

5.Chemical resistance.

6.Surface appearance

7.Electrical conductivity

8.Optical clarity in comparision to conventionally filled polymers.

7.9 Nano Fillers is the polymer matrix

Recent trend is to add nanofillers in the polymers matrix and Such type of composites are called as nano composite

Polymer + Nano filler = polymer Nano Composite

Example of nanofiller: Cellulose nano crystal/nanofibers

Carbon nano tube/ nanofibers.

Carbon block etc.

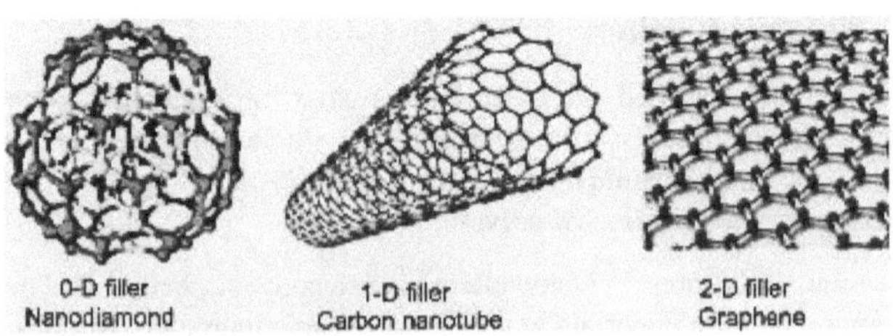

Fig. 7.1

are called nanofillers because dimension is in nano meter (nm). For example, Carbon nanotube's diameter is 5-20 nm. Because of their nano-Size, Surface area of these fillers is extremely high which is very important to achieve good reinforcing properties.

Polymers vary in their chemical Structure and offer different properties, therefore selection of polymer matrix made based on application and property requirement.

Nano Composite it can be polymer based Composite due to small size of the particles of the structural unit and high surface area to volume ratio nano material have large surface area but a very small volumes mean s is large and v is small (not talk gravitational forces talk on vander waal bonds surface tension , surface phenomena get into action to describe the process)[Where we talk about nano, gravity is no more important, density is no more no more Important].

7.10 Applications

Nanotechnology is deeply embedded in the designs of advanced devices for electronic and optoelectronic applications. The dimensional scale for electronic devices has now entered the nano range. The utility of polymer-based nanocomposites in these areas is quite diverse involving many potential applications and have been proposed for their use in various applications.

• They are used in chemical sensors, electro catalysis, batteries, Smart windows and memory devices.

• Another potential, application includes photovoltaic (PV) cells and photodiodes. Supercapacitors, printable conductors, light emitting diodes (LEDs)

• The electrical conductivity of carbon nanotubes in insulating polymers has also been a topic of considerable interest. The potential applications include electromagnetic interference shielding, transparent conductive coatings. electromechanical actuators.

• Conjugated polymers with various nanoscale filler inclusions have been investigated for sensor applications including gas sensors, blo sensor and chemical sensors. The nanofillers employed include metal oxide nanowires Carbon nanotubes, nanoscale gold, silver, copper, platinum .

• polymer based solar cells have the capability of being used to make cheap large flexible panels Cadmium chalcogenides such as CDS nanoparticles in polymers have been used to make solar cells.

More Books!

I want morebooks!

Buy your books fast and straightforward online - at one of world's fastest growing online book stores! Environmentally sound due to Print-on-Demand technologies.

Buy your books online at
www.morebooks.shop

Kaufen Sie Ihre Bücher schnell und unkompliziert online – auf einer der am schnellsten wachsenden Buchhandelsplattformen weltweit! Dank Print-On-Demand umwelt- und ressourcenschonend produzi ert.

Bücher schneller online kaufen
www.morebooks.shop

info@omniscriptum.com
www.omniscriptum.com

OMNIScriptum

FSC
www.fsc.org
MIX
Papier | Fördert
gute Waldnutzung
FSC® C083411

Kolibri 360 GmbH
Ferdinand-Jühlke-Straße 7
99095 Erfurt, Deutschland
produktsicherheit@kolibri360.de

Druck:
CPI Druckdienstleistungen GmbH
im Auftrag der
Kolibri 360 GmbH
Ferdinand-Jühlke-Str. 7
99095 Erfurt